Ecotoxicology

Ecotoxicology

F. RAMADE

Professor of Ecology and Zoology at the Université de Paris-Sud (Orsay)
Member of the IUCN Commission on Ecology (Gland, Switzerland)

Translated by L.J.M. Hodgson

JOHN WILEY & SONS
Chichester · New York · Brisbane · Toronto · Singapore

Copyright©Masson Editeur, Paris, 1979 2nd edition. First published February 1977

Copyright©1987 by John Wiley & Sons Ltd

All rights reserved.

No part of this book may be reproduced by any means, or transmitted, or translated into a machine language without the written permision of the publisher.

Library of Congress Cataloging in Publication Data:

Ramade, François.
  Ecotoxicology.
  Translation of: Ecotoxicologie.
  Bibliography: p.
  Includes index.
  1. Pollution—Environmental aspects. I. Title.
[DNLM: 1. Environmental Pollutants—adverse effects.
2. Environmental Pollution. WA 670 R165e]
QH545.A1R3513   1987   628.5   84-26999
ISBN 0 471 10445 0

British Library Cataloguing in Publication Data:

Ramade, F.
  Ecotoxicology.—2nd ed. rev. and enl./
  translated by L.J.M. Hodgson
  1. Pollution
  1. Title   II. Ecotoxicologie. *English*
  363.7′3   TD174
  ISBN 0 471 10445 0

Printed and bound in Great Britain

Contents

Preface to the 2nd (French) edition　　vii

Introduction　　ix

Chapter One—The concept of toxicity and its
　　　　　　　ecological implications　　1
A—Toxicants and toxicology　　1
B—Pathological problems peculiar to ecotoxicology　　22
C—The dose-response relationship in ecotoxicology　　31
D—The influence of ecological factors on the effects of toxicity　　42
E—Analytical methods for detecting pollutants　　52
F—The monitoring of pollutants　　54

Chapter Two—The pollution of the biosphere　　59
A—Pollution　　59
B—The causes and significance of the pollution of the biosphere　　61
C—The classification of pollutants　　70
D—Dispersion and circulation mechanisms of pollutants　　71

Chapter Three—Chemical pollution　　87
I—Pollutants affecting both continental and marine ecosystems　　88
A—Organohalogen compounds　　88
B—Mercury　　119
C—Cadmium　　131
II—Pollutants of continental ecosystems　　135
A—Pollutants of agroecosystems　　135
B—Atmospheric pollutants　　150
III—Pollutants of the hydrosphere　　181

Chapter Four—Nuclear pollution 203
A—Concepts of radiobiology 205
B—Ecological consequences of radioactive fallout 215
C—Radioecological consequences of the development of the
    nuclear industry 220

Conclusion 236

Bibliography 238

Subject Index 255

Taxonomic Index 260

Preface To The 2nd (French) Edition

Less than two years after its publication, the first edition of *Ecotoxicology* had already sold out, which indicates the increasing interest of the scientific community in the study of the ecological consequences of pollution. It is quite clear that during such a short period of time, the science of ecotoxicology has not evolved on a conceptual level.

Nevertheless, the reader will find in this second edition, which has been significantly enlarged in comparison to the first, many new developments regarding some of the issues, whose topicality has been demonstrated yet again by such recent catastrophic events as the shipwreck of the *Amoco Cadiz* or the accident concerning the nuclear reactor on Three Mile Island.

Among the numerous additions to the second edition, we should like to mention in particular the analysis of the problem of chemical carcinogens in the environment, and a study of the ecological consequences of pollution by cadmium, asbestos and dioxin (cf. the 'Seveso' affair.) The chapter on the pollution of the ocean by hydrocarbons has been expanded to include the analysis of several recent studies, notably the data provided by the research into the effects of the oil slick from the *Amoco Cadiz*. The chapter on nuclear pollution has been enlarged with an examination of the ways in which radionuclides become absorbed into food chains.

Finally, we have taken the opportunity of a second edition to correct several queries and printing errors which appeared in the first edition of the work.

Introduction

Unrecognized, if not deliberately ignored for far too long, the problems of pollution have quite recently become a constant preoccupation of the industrialized nations. During the last decade, countless meetings of experts, a large number of conferences and scientific congresses and a multitude of articles in the press have helped to popularize the notion of pollution and have also led to most people becoming aware of dangerous situations that had sometimes been in existence for years.

It would not, however, be true to suppose that this glut of information has as yet caused the governing politicians of the industrialized nations to take the radical, concrete decisions which alone could halt the dangerous route being taken by the civilized world. Proof of this is the hesitation, even the clear reluctance, with which the authorities in the so-called developed countries enforce legal restrictions on large companies that are a major source of industrial pollution.

In spite of this, scientific knowledge of the different forms of pollution of the biosphere and of its effects on all living things has made enormous progress over the last few years. This knowledge constitutes the necessary preamble to all measures aimed at fighting the contamination of our natural environment. At the same time, it is becoming increasingly evident that a good understanding of the mechanisms governing the action of pollutants cannot be achieved by using only *in vitro* experiments on organisms in the laboratory. The fact is that a huge variety of contaminating substances are deposited in the atmosphere, in the soil, in inland waters and in the oceans. Like natural substances, they too are subject to the interplay of biogeochemical phenomena. The transformations they may undergo in these different environments can of course neutralize them, but equally these environments can facilitate the dispersal of the pollutants and enhance their toxicity.

In addition, the contamination of the biosphere takes the form of indirect and delayed actions whose importance is greater than the effects of direct action on the biocoenoses. Polluting agents do not in any case attack isolated individuals in the natural world but affect whole populations and communities. Because of the immense number of interactions that occur within an ecosystem, imbalances can appear which are perhaps favourable to some species but harmful for most organisms, even if the particular pollutant introduced into a given habitat is not toxic for the majority of taxonomic groups making up the biotic community. Therefore, it is clearly essential that any investigation into the biological consequences of pollution must be viewed in its ecological context.

There is an increasing urgency for a synthesis to be made of all the work carried out over the past twenty years on the problems of contamination of the environment, and to gather these studies together under the name of an emerging specialist science, viz., ecotoxicology. It can be defined broadly as the science whose aim is to study the different forms of environmental contamination by both natural pollutants and artificial ones produced by human activities, as well as to study the mechanisms of their actions and their effects on all living organisms in the biosphere. The science of ecotoxicology, like most other sciences, has a fundamental basis and a vast range of applications. It gives the opportunity not only for evaluating the importance of the effects on different ecosystems resulting from their contamination, but also for predicting to a certain extent future consequences of the release of a specified pollutant.

The pollution of the biosphere by our technological civilization does not just threaten the long-term survival of the animal and plant species that inhabit the Earth. It is also jeopardizing the future of humanity itself by gradually squandering irreplaceable natural resources, in particular those which govern the agricultural productivity of the different continental ecosystems. Furthermore, the dispersal of toxic substances in the natural environment leads to contamination of human food chains, which is even more dangerous given that our species is in every respect at the summit of the ecological pyramid. Humanity's thoughtless behaviour in disposing of all the waste substances of our activities into the environment, exposes us to a sort of 'boomerang effect' by means of food chains. Pollution of the oceans or of the atmosphere can have harmful effects not just directly, when man is subjected to a contaminated environment, but more subtly by polluting the trophic network. This has already caused dramatic events such as the famous Minamata disease in Japan.

Ecotoxicology, in order to be effective, has to acquire a profound knowledge of the fundamental concepts of ecology as well as of the principal physiological mechanisms governing the action of pollutants. Any rational approach to this science also necessitates successive studies on the effects of toxic substances on an individual (the ecophysiological level), on a population (the 'demoecological' level) and on an ecosystem (the synecological level). This is also the basic approach adopted by the book.

CHAPTER ONE

The Concept of Toxicity and its Ecological Implications

A—Toxicants and toxicology	1	C—The dose—response relationship in ecotoxicology	31
I—Ways in which toxicants penetrate an organism	2	1. Dose accumulation and genotoxic effects	32
II—Different effects of toxicity and their evaluation	3	2. The notion of the maximum admittable concentration	37
1. Effects of toxicity	3		
2. Evaluation of the toxicity of a substance	4	D—The influence of ecological factors on the effects of toxicity	42
III—Principal types of physiotoxicological effects	13		
1. Principal somatic changes	13	I—The influence of intrinsic factors	42
2. Principal germinal effects	21	II—The role of extrinsic factors	49
B—Pathological problems peculiar to ecotoxicology	22	E—Analytical methods for detecting pollutants	52
1. Permanent exposure	22		
2. Consequences	23	F—The monitoring of pollutants	54

A—TOXICANTS AND TOXICOLOGY

The concept of 'toxic', though familiar, has in reality a wide variety of meanings. To the layman, 'toxic' is a synonym of 'poisonous'. The scope of ecotoxicology does in fact overlap very broadly with the specialized study of poisons, which can be defined as chemical substances possessing high levels of toxicity for mammals and man. A broad definition of this term will be used in the book and applied to any physical (heat, radiation), chemical or biological factor that creates a potential source of pollution.

Even if the reader is familiar with the basic concepts of ecology, he is probably far less so with those of toxicology, which is a much narrower field of specialization. The aim of toxicology is to study the different problems relating to toxicants on an analytical level as well as on the physiological and biochemical levels. It is both a descriptive and an explanatory science in the way in which it seeks to define accurately the mechanisms governing the actions of poisons. It covers all analytical research into toxic chemical substances present in different environments or inside living organisms. But the term 'toxicology' is also given to all investigations designed to evaluate the toxicity of pollutants on living species. Truhaut (1974) defines toxicology as 'the science which studies toxic substances or poisons, that is substances which cause alterations or perturbations in the functions of an organism leading to harmful effects, of which evidently the most serious is the death of the organism in question'.

Other authors give a much more restricted definition of the field of toxicology. For instance, O'Brien (1967) reserves the use of this term for the study of 'the mechanisms by which toxic substances exercise their effects'. Without going so far as to adopt O'Brien's definition, it can be proposed that the analytical aspects of toxicology and even the evaluations of immediate toxic effects constitute the technical methods of approach of this science, all of which are of course essential, but which do not really highlight the most important aspect of the science, namely the study of mechanisms governing the action of toxicants on the molecular and cellular levels as well as on entire organisms.

Similarly, the most specific function of ecotoxicology relates to the study of the ways in which polluting agents disturb populations and communities, and is not just concerned with detecting traces of a particular substance contaminating a given environment.

Until quite recently, toxicology had devoted itself almost exclusively to very toxic substances whose sphere of activity is towards the higher levels of toxic concentration, or above 25 mg/kg of the body weight of a living organism. Nevertheless, it had been recognized for many years that certain toxic compounds exist which, when present in the biosphere in normal amounts, are harmless or even necessary for living beings. For example, carbon dioxide, oxygen or sodium chloride, in certain concentrations, can cause serious disorders in both autotrophic and heterotrophic organisms. These facts bear out the ancient proverb of Paracelsus *sola dosis fecit venenum* (only the dose makes the poison). All the same, it was not until after the Second World War that, due to the proliferation of synthetic organic products and other polluting agents of various kinds, the science of toxicology was forced to take a serious interest in substances which were formerly thought of as non-toxic and whose toxic potential even appeared to be ridiculously low.

I — Ways in which Toxicants Penetrate an Organism

At the present time, there are a multitude of toxic substances in the soil, the air, water and foodstuffs. If, in the majority of cases, the contamination of the environment and of food for domestic animals and man is unintentional, there are, however, many cases where it is almost deliberate: for example, in the use of pesticides and of certain food additives.

In toxicology, three standard categories are defined for means of absorption (or three methods of contamination):

1. Respiratory intake—this is the most common method of contamination by atmospheric pollutants.
2. Transcutaneous intake.
3. Ingestive intake (absorption by the roots of plants or by the digestive system of animals).

In the animal world, the following forms of toxicity correspond to the above methods of penetration: (1) by inhalation; (2) by contact; (3) by ingestion (or orally = *per os*).

Similarly, plant life is exposed to these three methods of contamination: by the direct diffusion of toxic gases through the epidermis of the leaf; by stomatic respiration, where atmospheric pollution is concerned; and by absorption through the roots, when there is pollution in the soil.

For aquatic organisms it is impossible to distinguish different methods of absorption such as percutaneous from ingestion, as they occur simultaneously.

II—Different Effects of Toxicity and their Evaluation

1. Effects of toxicity

Living organisms can show different physiological reactions to the same toxic substance, depending on the quantities absorbed and the exposure time.

Acute Toxicity

This provokes the most drastic reaction to the toxicity of a poison. It is characterized by the rapid death of the contaminated individual or population, and leads the layman to think that any substance which kills violently is poisonous.

Acute toxicity can, therefore, be defined as causing death or very serious physiological disorders shortly after absorption through the skin, by the lungs or by mouth of a fairly large single or repeated dose of a poisonous compound.

The inhalation of carbon monoxide, the absorption of parathion or cyanide in low doses taken in absolute value (approx. 10 ppm) are good illustrations of the spectacular aspect of this form of toxicity.

Sublethal Toxicity

This differs from the acute form in that a significant proportion of a population can survive exposure to the toxicant, although each individual shows a clinical reaction resulting from its uptake.

Long-term Toxicity

Research in ecotoxicology is principally concerned with toxic effects produced not by the absorption of fairly strong doses over a short period, but with exposure to very low concentrations, even sometimes in minute doses, of pollutants whose repeated cumulative effects result in far more dangerous perturbations.

As Truhaut (1974) stresses, the term 'chronic exposure', which is often used to describe this type of effect, is incorrect: irreversible and, therefore, chronic damage can in fact result from an initial occurrence of acute toxicity.

2. Evaluation of the toxicity of a substance

Toxicological Tests

The aim of these tests is to evaluate the degree of sensitivity (or resistance) of different animal and plant species to a particular toxic substance. In practice, it involves determining the different forms of toxicity: by inhalation, by contact or by ingestion. In animal toxicology, the methods most generally used are subcutaneous or intraperitoneal injections, allowing stricter control of the doses administered.

The toxic potential of a substance, whether its effects are acute or long term, is calculated by establishing various parameters that characterize its action not on an isolated individual but on a population. The most serious consequence of pollution—the death of the contaminated organisms—can only be estimated by means of a mortality rate (or coefficient) based on the reaction of the whole population rather than of an individual. It represents, therefore, a demo-ecological criterion. Estimating the mortality rate caused by a toxic substance under standardized experimental conditions on a sample from a population of a reference species makes it possible to evaluate the different forms of toxicity of a given substance.

In general, three essential precautions must be taken when doing any toxicological test:

1. The sample of subjects from the test species being used in the experiment should be as homogeneous as possible by selecting individuals of the same sex, age and weight.
2. The technique for administering the toxicant should be one which guarantees constant experimental conditions throughout the test.
3. Data from the experiments should be collected selectively and analysed using an appropriate statistical method.

Principal Toxicological Parameters and their Definition

The aim of toxicological tests is to calculate the principal parameters characterizing the acute or long-term toxicity of any substance thought to be poisonous.

(a) Calculation methods—There are two which can be used:

The first involves determining the mortality rate y after a fixed time (for example, 24 hours or 3 months) as a function of increasing doses of the toxicant: $x_1, x_2, x_3, \ldots x_n$. From this can be drawn the representative curve of the function $y = f(x)$. The different characteristic constants of the compound being studied can then be calculated. The most important constant which can be determined by using dose–mortality tests is the LD_{50} (= median lethal dose), also called the 50% lethal dose. For volatile compounds (with variable concentration levels in the air), or for compounds tested on aquatic organisms, the use of these dose–mortality tests can determine the LC_{50} (= median lethal concentration), a theoretical value like the LD_{50}, causing 50% mortality in the population being studied. The LC_{50} is also referred to in tests on compounds that act by contact. Then the concentration is expressed in terms of units of area and not of volume (for example in the case of a substance dispersed over the leaf area of a plant).

With certain substances and in certain species of invertebrates it may be difficult to define mortality. In such instances, reference is made to another parameter, no longer the lethal dose, but the immobilizing concentration (IC). The IC_{50} corresponds to the concentration inhibiting the mobility of 50% of the test population. This parameter is frequently used in tests on aquatic organisms. A variant of it widely used in insecticide tests is based on the 'knock-down' principle. This reaction is characterized by motory ataxia or incoordination following exposure to the toxicant, with the affected insects unable to fly and lying paralysed on their backs. The 'knock-down' state can also have a median, or $k.d._{50}$.

A final parameter frequently used in ecotoxicology is the EC_{50}, the effective concentration needed to provoke a given sublethal response. This parameter is often employed to determine the 'tolerable' level of concentration of a pollutant by showing the variation in the EC_{50} as a function of time.

By extension, dose–mortality tests can be used to calculate both the LD_{10}, which makes the boundary between acute toxicity and sublethal toxicity (i.e., mortality less than 10%), and the LD_{90}, which has an obvious practical application when high toxicity is purposely investigated (the 'screening' of pesticides, for example). The LD_{50} and LC_{50} can usually be determined after 24 hours or 48 hours of exposure in tests of acute toxicity, and in some cases after 96 hours.

A second method of testing toxicity involves determining the mortality rate y following the application of a constant dose as a function of increasing periods of time $t_1, t_2, t_3, \ldots t_n$ from which can be established the equation $y = f(t)$. It is then possible, by using this equation, to calculate the LT_{50} (= median lethal time), the theoretical length of time it would take for the death of 50% of the individuals exposed to a given dose (or concentration).

The absolute value of the preceding parameters can vary immensely for different species according to the conditions of exposure which characterize the type of test chosen. Nevertheless, these tests make it possible to obtain a series of experimental values which express the dose–mortality or time–mortality relationships.

(b) The transformation of the probit—The numerical values obtained can be used to draw empirically the approximate graphs from which one can try to define the relationship between E(y), the mathematical expectation of mortality rate, the dose x (with a fixed time) or the time t (with a fixed dose). Empirical graphs could never be used, without the risk of gross errors, to draw valid conclusions on the quantitative level, nor to make a comparison between different experiments. Accurate exploitation of the results calls for an appropriate statistical method which can on the one hand produce the most likely graph from the experimental data, and on the other hand enable an evaluation to be made of the significance of the test. This can be done by using the transformation of the profit.

The application of this statistical method to toxicology was introduced by Gaddum (1933) and Bliss (1935). Subsequently Sheppard (1947) and then Hoskins (1947–60) adapted it to tests on pesticides.

The object of this transformation is to mitigate the fact that the direct representation, on a mathematical scale, of mortality percentages y_i on the y-axis and doses x_i on the x-axis produces curves of the sigmoid type which are difficult to manipulate and do not lend themselves well to an estimation of the LD_{50}. Transformations have, therefore, been sought of both variable and function $z = g(y)$ and $u = h(x)$, such as $z = k(u)$, that is a function whose representative curve is the simplest possible, preferably a straight line.

(c) Principle of the transformation of the probit—The transformation of the probit is based on the hypothesis that if one takes a fairly large and homogeneous population, the frequency of individual minimum lethal doses (that is, the weakest dose capable of killing each isolated individual, which of course varies from one individual to the next) is distributed over a normal variable density curve well known by biologists under the name of a Gauss bell curve (Fig. 1.1). This can be expressed by saying that the mathematical expectation of the proportion of deaths E(y) and the dose x are related thus:

$$x \to E(y) = \pi \frac{(x - \mu)}{\sigma} \tag{1}$$

where π is the distribution function of the transformed normal rule.

The representative curve of the points x, E(y) is called a sigmoid. If we replace E(y) by the variable z defined as $\pi(z) = E(y)$ we can express this as:

$$x \to E(y) = \pi \frac{(x - \mu)}{\sigma} = \pi(z) \tag{2}$$

As the function π is strictly increasing and, therefore, never has the same value twice, we now have:

$$z = \frac{(x-\mu)}{\sigma} \qquad (3)$$

Numerous studies have confirmed that toxic substances, after transformation of the probit, led to the linear relations $z = k(u)$ in which u is a function of x (Fig. 1.1).

The theory of linear regression, therefore, enables the straight line to be calculated which best fits the data coordinates (μ_i, z_i).

In order to avoid negative signs for the proportion of deaths less than 0.5 Bliss added 5 to the standard deviations of the transformed normal variable. Under these conditions, the probits that are less than 5 correspond to a mortality rate of less than 50%.

A further difficulty has arisen in toxicology owing to the fact that the individual lethal doses do not follow the normal law. Earlier studies showed on the other hand that the variable $u = h(x)$ where h is the logarithm function leads to the relation:

$$u \rightarrow E(y) = \pi \frac{(x-\mu)}{\sigma} \qquad (4)$$

The graphic representation of toxicological tests implies, therefore, a double transformation of both variable and function, the time and the doses being replaced by their logarithm and the proportion of deaths by their probit.

In this way, the linear functions $u \rightarrow z$ are obtained whose representative straight lines were called by Hoskins ld-p lines (logarithm of the dose-probit) and lt-p lines (logarithm of the time-probit), which make it possible to calculate the different parameters characterizing the action of a toxic substance on any species of animal or plant.

(d) Calculation of the ld-p and lt-p lines—The techniques of linear regression enable the straight line to be drawn which best fits the transformed experimental points and to calculate its equation. It should be noted that the very nature of toxicological tests means that a relatively simple form of regression analysis can be adopted, because one of the variables, the dose or the time being supposedly controlled in the experiment, loses its character of a random variable so that only the regression of the one random variable (the mortality percentage transformed to a probit) need be taken into account in relation to the controlled variable.

Generally in a linear regression it is shown that the regression coefficient *by/x* is easily calculated from the n pairs of experimental values (cf., for example, Lamotte M., *Initiation aux methodes statistiques en biologie*. Masson, 1962),

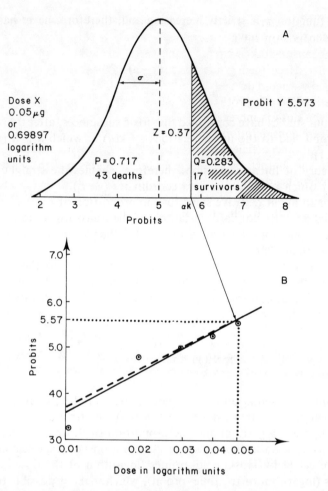

Fig. 1.1 Principles of the transformation of the probit Fig. 1.1.A represents the distribution curve of the individual lethal doses and Fig. 1.1B shows the double change of variable and function enabling the transformation of the distribution curve or sigmoïd for the proportion of deaths (or mortality percentages), into the straight line $ld-p$. The experiment shows that in changing the variable $u = h(x)$ where u is the logarithm function, the frequency of individual lethal doses is distributed over a normal variable density curve (well known by biologists as a Gauss bell curve). For every experimental dose x_k there corresponds a point on this normal curve of the abscissa y_k so that a parallel line to the y-axis drawn through this point, the curve and the x-axis describes an area which represents the observed proportion of deaths, an integral of the normal curve between $-\infty$ and α_k. The total area between the curve and the x-axis represents the total number of the population. This area is simply the value of the function whose representative curve is a sigmoïd. The probit corresponding to the observed proportion of deaths, y_k has the value $z = \alpha_k + 5$, Bliss (1935) having added 5 to the standard deviations in order to avoid negative signs for the mortality percentages less than 50%. z corresponds as it were to a theoretical dose assumed to cause in the population the proportion of deaths y_k, being identical to the proportion determined by the dose x_k actually used. The data chosen to illustrate this figure are from a test made on a strain of

but at this point there occurs an additional difficulty in that the variance of a probit depends on the proportion of deaths. It can be seen from this that the probits must be taken into account when calculating the regression line, as the error that affects them is minimal for $y = 5$ (50% mortality rate) and increases symmetrically when n deviates from this value. Likewise, it is logical to expect that the estimation of a probit will be more reliable as the number of test subjects for the given dose is increased.

All of these considerations lead to the introduction of an auxiliary probit y_w and a correlation coefficient w by which the probit must be multiplied in order to calculate the characteristic parameters of the regression line. Fisher's theory of the maximum probability can be used to show that the auxiliary probit y_w is linked to the probit Y, which is deduced directly from the proportions of experimental deaths by the expressions:

$$y_w = Y + \frac{Q}{Z} - \frac{s}{n} \times \frac{1}{Z} = Y + \frac{Q}{Z} - \frac{q}{Z} \qquad (5) \quad (\text{if } Y > 5)$$

$$y_w = Y - \frac{P}{Z} + \frac{n-s}{n} \times \frac{1}{Z} = Y - \frac{P}{Z} + \frac{p}{Z} \qquad (5') \quad (\text{if } Y < 5)$$

In these equations P and Q are the probabilities of death and survival for the tested doses corresponding to the probit Y, Z is the ordinate of the point on the curve corresponding to the probit Y, and p and q are the proportions of deaths and survivors actually observed for this dose with s as the number of survivors for n numbers of individuals tested.

It is also necessary for the auxiliary probit, corresponding to each dose, to have a correlation coefficient, which from the theory is given the value:

$$W = N \frac{Z^2}{PQ}$$

The parameters y, Y, $Y + \frac{Q}{Z}$, $Y - \frac{P}{Z}$, $\frac{1}{Z}$ and $\frac{Z^2}{PQ}$

Fig. 1.1 (*continued*) house-flies. A dose of 0.05 μg of an insecticide, lindane, caused the death of 43 flies out of the 60 treated, that is a 71.7% mortality rate (or a proportion of deaths of 0.717). This proportion corresponds to the blank area between the normal curve and the x-axis, described by a parallel line to the y-axis drawn through the abscissa point $z = 5.573$, a probit corresponding to the percentage of observed deaths. The arrow relating this point to Fig. 1.1B symbolizes the transformation of the probit. In effect, the value $z = 5.573$ is taken to the y-axis of the transformed curve in place of the mortality percentage, while on the x-axis the dose is replaced by its logarithm (0.6988, the dose having been multiplied by 100 to avoid a negative characteristic). (After Ramade, 1967.)

can be found in the tables of Fisher and Yates (Tables IX-1 and 2, p. 71; Oliver and Boyd, 1963).

The first of these tables provides the y probits, corresponding to the proportion of observed deaths, and the second indicates the other parameters from the theoretical probit Y. Nevertheless, the first step when beginning to use Table IX-2 must be to determine the theoretical probit Y from the experimental points, which differs from the experimental probit y. It is for this reason that an empirical straight line (or provisional straight line) is drawn by eye, so that the necessary hypotheses can be obtained to use Fisher's tables. All that is needed is to read directly from this straight line, by a graphic evaluation, the (rough) values corresponding to each dose (Fig. 1.2).

(e) Graph of the empirical straight regression line—This can be drawn from the experimental data by determining by eye the line which best seems to fit the various points on the graph. It is most important to draw this line with great care, as later calculations depend on its accuracy. The line on the graph is obtained by using the method of least squares.

Firstly, the median point of the straight regression line is calculated, whose

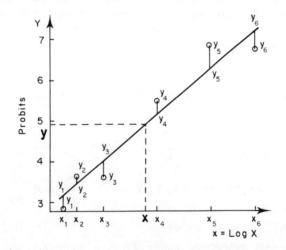

Fig. 1.2 Graph by eye of the empirical straight regression line in the processing of experimental data obtained from a test of toxic substances using the probit method. The best straight line that can be drawn by eye is one where the sum of the distances from the coordinates below the line to the line drawn is equal to the sum of the distances from the coordinates above the line to the line drawn. This empirical straight line has to pass through the median point of the coordinates (X,Y).
(After Ramade, 1967)

coordinates are:

$$\bar{X} = \frac{\Sigma x_k}{n} \text{ et } \bar{Y} = \frac{\Sigma y_k}{n}$$

where n is the number of pairs of values. A ruler is then pivoted round this point until the sum of the distances of the points from above and below the ruler meet the following conditions:

$$\Sigma \overline{yy}' = \Sigma \overline{yy}''$$

Once the line is drawn, it is possible to read off the theoretical probit corresponding to each dose.

The equation of the theoretical regression line is calculated using the same principles as linear regressions by taking into account each time the coefficient w (which is the equivalent of w number of observations, each observation having a unit weight). In this way, the coordinates of the median point become:

$$\begin{cases} \hat{y}_w = \dfrac{\Sigma(wy_w)}{\Sigma w} \\[2ex] \hat{x} = \dfrac{\Sigma(wx)}{\Sigma w} \end{cases} \qquad (6)$$

The equation of the regression line is expressed, by analogy with the unweighted regression, thus:

$$Y = \hat{y}_w + b(x - \hat{x}) \qquad (7)$$

where

$$b = \frac{\Sigma[wy_w(x - \hat{x})]}{\Sigma[w(x - \hat{x})^2]} \qquad (8)$$

The sums of the squares and the products can be calculated more rapidly if their expression is transformed to:

$$\Sigma[w(x - \hat{x})^2] = \Sigma(wx^2) - \frac{\Sigma^2(wx)}{\Sigma w} \qquad (9)$$

likewise

$$\Sigma[w(y_w - \hat{y}_w)^2] = \Sigma(wy_w^2) - \frac{\Sigma^2(wy_w)}{\Sigma w} \qquad (9')$$

and

$$\Sigma [wy_w (x-\hat{x})] = \Sigma(wxy_w) - \frac{\Sigma(wx)\Sigma(wy_w)}{\Sigma w} \qquad (9'')$$

Having established the equation of the regression line, it must be checked to ensure it represents the data well. To do this, the following expression:

$$\Sigma [w (y_w - \hat{y}_w)^2] \qquad (8')$$

the sum of the squares of the deviations of the probits from the theoretical median, must be compared with:

$$\frac{\Sigma^2 wy_w (x-\hat{x})}{\Sigma [w (x-\hat{x})^2]} \qquad (10)$$

the quantity from which the sum of the squares of yw is reduced.

It can be shown that these expressions, just as their difference or residue, are all χ^2. The straight line drawn is even more precise when the difference between the expressions (8') and (10) is small. In principle, this difference, or χ^2 of the error (residue) has n degrees of freedom where $n = N - 2$ with $N =$ the number of doses tested, when verifying the corresponding probability in the table of χ^2.

It should be noted, however, that in the practice of toxicological tests, a high χ^2 does not mean the straight line calculated has to be rejected if, following new hypotheses, a second or a third calculation does not obtain more reduced residues. Such cases simply show that the population being studied is heterogeneous, due either to problematical variations (size, age or sex) or to the presence simultaneously of individuals who are sensitive to the toxic substance and those who have a resistance to it.

(f) Calculation of the LD_{50} and the standard deviation—A probit $Y = 5$ corresponds to the dose causing a 50% mortality rate. From equation (7) it is easy to deduce:

$$\text{Log } LD_{50} = \hat{x} + \frac{5 - y_w}{b} \qquad (11)$$

Fisher's theory on estimation and the maximum of likelihood shows that the amount of information (Ix for the doses) varies inversely to the variance. Thus, in a weighted regression line:

$$I_x = I_y \left(\frac{d_y}{dx}\right)^2 \text{ and } V_x = V_y \left(\frac{dx}{dy}\right)^2 \qquad (12) \text{ and } (12')$$

as dy/dx is the same as b, the slope of the straight regression line,

$$V_x = \frac{1}{b^2} V_y,$$

which expresses the variation of the probit as a function of the dose.
By replacing V_y in its expression, it can further be deduced:

$$V_x = \frac{1}{b^2} \left[\frac{1}{\Sigma w} + \frac{(x - \hat{x})^2}{\Sigma [w(x - \hat{x})^2]} \right] \qquad (13)$$

from which comes the variance in the LD_{50}

$$VLD_{50} = \frac{1}{b^2} \left[\frac{1}{\Sigma w} + \frac{(\text{Log } LD50 - \hat{x})^2}{\Sigma [w(x - \hat{x})^2]} \right] \qquad (14)$$

from which the standard deviation in the LD_{50}

$$\sigma LD_{50} = \sqrt{V_{LD50}} \qquad (15)$$

III — Principal Types of Physiotoxicological Effects

The ways in which toxic substances act on living beings can be divided into two groups:

1. The first group are the **somatic effects**, or those which affect one or more functions of vegetative life or any related functions — whether they arise from acute, chronic or long-term toxicity.
2. The second group, **germinal effects**, concern all perturbations of the reproductive functions of contaminated individuals or any action which affects the physical entirety of their descendants through teratogenic or mutagenic effects.

The action of toxic substances can be translated into a large variety of physiotoxicological effects. However, the toxic phenomena most likely to cause rapid death are generally poisons that act on the nervous system (anticholinesterase compounds, for example) or that inhibit cellular respiration.

1. Principal somatic changes

In spite of the enormous variety of physiological perturbations caused by toxic substances, it is possible to recognize a few predominant physiotoxicological effects.

Neurotoxicity

Without doubt, this presents the most spectacular form of poisoning, and from it result the majority of cases of acute poisoning. The strongest poison or 'nerve gases' of chemical warfare (Sarin, Tabun, Soman, Vx, etc.) are neurotoxic agents, as is the famous botulinic toxin extracted from *Clostridium botulinum*.

The main reason for the extreme sensitivity of animals to neurotoxic substances lies in the essential role of the nervous system in the control of various vegetative functions. In addition, nerve cells are hypersensitive to any interference, however brief, in their metabolism, and any damage is irreversible.

The death of an individual often occurs in the final analysis as a result of nerve damage even in the case of poisons whose primary effects are on other organs. This is true of cardiotoxic substances such as atropine or substances that affect respiratory functions, such as carbon monoxide and cyanide, which finally cause death by the anoxia or deficiency of oxygen to the neurones, whose oxygen requirements are very high, thus causing irreversible cerebral damage.

Understanding neurotoxicological phenomena requires some detailed knowledge of the physiology of the normal nervous system.

(a) Nerve conduction—This involves two distinct processes: the transmission of information in the form of a nerve impulse along a specialized cell, the neurone (called axonal transmission); and the transmission of information from one nerve cell to the next through special zones of contact, the synapses (called synaptic transmission).

The nerve impulse consists of a depolarized wave transmitted in one direction only, and called an action potential. This transitory phenomenon, which involves

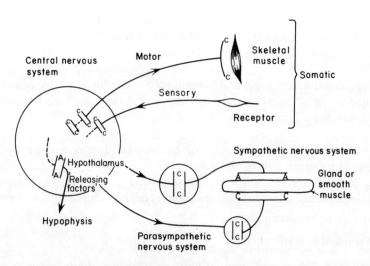

Fig. 1.3 A model of the organization of the nervous system of vertebrates, showing the main locations of the cholinergic and adrenergic synapses.
C = cholinergic   A = adrenergic

a refractory period, results from a modification of the tonic balance between the inner and outer axonal compartments.

Axonal transmission is the result of a process of active transport of ions across the neurilemma. When a depolarized wave has travelled through a neurone it arrives in the synaptic region where it can induce the release into the synaptic gap of a special substance (the mediator of nerve transmission, still called the neurotransmitter) if the action potential reaches a high enough level. This substance will momentarily modify the permeability of the post-synaptic membrane to ions, thus causing the genesis of a new action potential in the next neurone. In the case of effector neurones (i.e., neurones which command a particular organ) there can be synaptic contact of the axon-muscle type or axon-secretory gland type.

The principal chemical mediators of nerve transmission are acetylcholine, various catecholamines (e.g., noradrenaline, adrenaline, dopamine) and the 5-hydroxytryptamine (serotonin). The three types of neurones corresponding to the different chemical neurotransmitters are called cholinergic, adrenergic (in the widest sense) and tryptaminergic respectively.

Synaptic transmission involves enzyme systems capable of rapidly deactivating the mediator that has been released into the synaptic region immediately after its release. If this fails to happen, there will be a blockage of the synapse due to a real electrochemical short circuit. The deactivation of neurotransmitters is the job of specialized enzymes (e.g., acetylcholinesterase, monoamine oxidase) associated with the synaptic membranes. However, other deactivation processes seem to intervene in the case of adrenergic synapses.

(b) The action of neurotropic toxins — A large number of neurotropic agents act by blocking axonal transmission. This is the case with DDT, the contact insecticide whose method of action at the cellular level was long misunderstood. DDT disrupts the transfer of the nerve impulse by inhibiting the K^+- and Ca^{2+}-ATPase which control the active transfer of ions through the neurilemma.

Various natural toxins such as tetrodotoxin or the venom of the black widow spider (*Lathrodectes 4-guttatus*) cause irreversible paralysis by disrupting the transfer of ions through the neurone membranes.

A large number of neurotoxic agents block synaptic transmission at the cholinergic synapses. Organophosphorus insecticides, just as all phosphoric esters produced synthetically, are powerful inhibitors of acetylcholinesterase enzymes. They act by settling on the active sites of the enzyme, inhibiting it irreversibly (cf. Fig. 1.4). Various natural carbamates (eserine, prostigmine) or synthetic carbamates (insecticides such as Sevin or Baygon) behave like inhibitors in competition with the cholinesterases by settling on active sites of the enzyme, but this is reversible inhibition.

Most synaptic poisons act on the cholinergic junctions. However, the response of the cholinergic synapses is not uniform but varies according to the toxic substance used. A first group of cholinergic synapses in vertebrates comprises those of the parasympathetic ganglia and the neuromuscular junctions. These are sensitive to nicotine, whose effects are characterized by paralysis, fasciculation

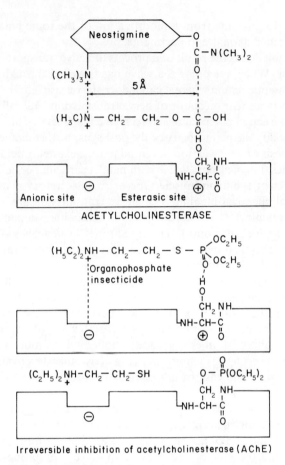

Fig. 1.4 The mode of action of anticholinesterasic toxic substances (schematized). The two active sites of acetylcholinesterase are separated by a distance of 5 Å. The same distance separates the grouping of quaternary ammonium carboxyl in the substrate of this enzyme, acetylcholine. All the inhibitors of acetylcholinesterase possess likewise two groups, one electropositive and the other electronegative, separated by a distance of 5 Å which, therefore, allows them to bond perfectly with the anionic and esterasic sites of the enzyme.

In the case of competing inhibitors (for example, neostigmine—a natural carbamate), the bonding is reversible after the hydrolysis of the substance. On the other hand, with an irreversible inhibitor such as amiton for example, which is an organophosphorus insecticide (see the bottom diagram), the esterasic site is completely blocked after the inhibitor is broken down by hydrolysis

(trembling) of the striated muscles and stimulation of the ganglia of the vegetative nervous system (called the nicotinic syndrome). A second group of cholinergic synapses in vertebrates are sensitive to muscarine. These synapses are situated in the cerebral cortex and in the peripheral synapses of the parasympathetic nervous system. The toxic chemicals which interfere with these areas cause a syndrome called the muscarinic syndrome, which is characterized by psychic effects (hallucinations), a reduction in the heart rate, contraction of the pupils,

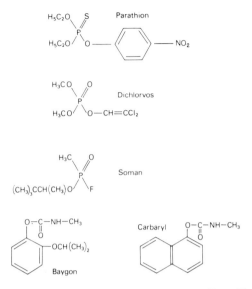

Fig. 1.5 Various anticholinesterasic agents. Parathion and dichlorvos (known popularly under the name Vapona®) are organophosphorus insecticides. Notice the similarity of their structure to that of Soman, a strong 'nerve-gas'. Baygon and Carbaryl are *N*-methylcarbamate insecticides which behave as reversible inhibitors of cholinesterases

urination, hypersalivation, etc. Neurotoxic agents affecting the cholinergic synapses are called nicotine or muscarine types according to their effects. Some of these substances can in fact produce both groups of symptoms simultaneously.

Other toxic substances act electively on the neurones of the adrenergic and/or tryptaminergic type. This is the case with a large number of neurotropic drugs such as the sedatives, (chlorpromazine, for example), the hallucinogens (LSD 25), the amphetamines, etc. All of these substances have structural similarities to the mediators of those groups. Other neurotropic drugs (nialamide) and even some insecticides (chlordimeform) disrupt adrenergic conduction by inhibiting monoamine oxidase.

A certain number of substances can cause degeneration of nerve tissue after chronic or long-term exposure. For instance, absorption of certain organophosphorus compounds such as tri-*ortho*-cresyl-phosphate (TOCP) or di-isopropyl-fluoro-phosphate (DFP) cause nerve degeneration in certain warm-blooded vertebrates by progressive demyelination of the axonal fibres.

Elsewhere, it has been possible to observe the damage done to the cerebral cortex, with degeneration of the perikaryon, caused in laboratory animals fed permanently on foodstuffs contaminated by organic mercury compounds or organochlorine insecticides (dieldrin, for example).

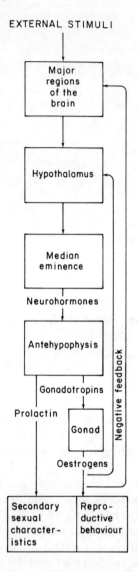

Fig. 1.6 Model of the neuroendocrinal system in vertebrates. Toxic substances can cause endocrine disturbances by acting on the central nervous system or by modifying the hormonal balance through interference with metabolism (modified negative feedback). (From *Pesticides and the Reproduction of Birds*, David B. Peakall. *Copyright© 1970 by Scientific American, Inc. All rights reserved.*)

Disruption of the Endocrinal Balance

Because of the importance of the neuroendocrinal balance, a number of toxic substances affecting the nervous system also act on the hormone balance of the organism. Some examples of the main problems caused by chronic or long-term exposure are the adrenal lesions and thyroid malfunctions which occur after absorption of various insecticides, and also the disturbance of reproductive functions caused by organohalogen compounds.

The sterilization of many species of birds, particularly birds of prey and ichthyophagous (fish-eating) birds, due to organochlorine insecticides, is the result of damage to the endocrinal balance. The sexual hormone balance is affected and this alteration to hormone regulation causes a reduction in activity in the gonads and a whole series of related pathological disorders, such as thinning of eggshells.

Damage to Respiratory Functions

Certain poisons act fairly selectively, some on cellular respiration, and some on the lung itself. Examples of inhibitors of cellular respiration are the cyanide compounds, which inhibit the cytochrome oxidase system, or various arsenical compounds such as Lewisite ($ClCH = CHAsCl_2$) which inhibits the pyruvate oxidase system. Other arsenic derivatives interfere with various other sulphydryl enzymes associated with the Lynen helix or the Krebs cycle. In vertebrates, the respiratory function can be blocked at the point when oxygen combines with haemoglobin (pollution by carbon monoxide (CO) causes the formation of carboxyhaemoglobin).

Various toxic substances can cause more or less irreversible damage to the respiratory organs through chronic or long-term exposure. In industry there are many strongly reactive chemical compounds which can destroy pulmonary tissue following accidental or long-term exposure. Even substances taken orally (*per os*) can cause irreversible pulmonary lesions. There have been some very strange cases of pulmonary fibrosis pointed out recently which were caused by accidental *ingestion* (and not inhalation) of a herbicide, paraquat.

Among the varied dangerous effects on the respiratory function due to the absorption of micropollutants there appears a recent and worrying problem — that of asbestosis. This serious illness occurs as a result of inhaling asbestos particles, a material used extensively in building and technical engineering (clutches, brakes). The illness takes the form of a pulmonary mesothelioma which is irreversible and fatal.

Another example is the constant rise in cases of chronic bronchitis which results partly from the atmospheric pollution caused by the release of sulphur dioxide from the burning of domestic and industrial fossil fuels. However, smoking habit is by far the main cause of this illness. The synergistic effects of sulphur dioxide and tobacco smoke are in fact discussed in Chapter 3.

Damage to Excretory Organs

The principal secretory organs, the liver and the kidneys, are particularly prone to serious organic lesions after chronic or long-term exposure to a pollutant.

(a) Hepatic lesions — One of the better-known disorders is the hepatotoxic action of the phalloidine, and another is the action of organic mercury compounds on the kidneys. However, these organs can also undergo serious

changes when exposed permanently to various micropollutants. A frequent reaction of the hepatocytes (hepatic cells) to permanent contamination of the organism by a given toxic agent takes the form of a steatosis (fatty degeneration)—a cytoplasmic surfeit in the lipid vacuoles. This occurs when large doses of certain compounds are absorbed—ethylic cirrhosis is an extreme example—but also after prolonged exposure to various micropollutants. For instance, a steatosis of the hepatocytes can occur in laboratory rodents fed experimentally on food contaminated by weak doses of organochlorine insecticides sometimes lower than 1ppm.

The absorption of medicinal substances, various pollutants or other toxic compounds causes a proliferation of the endoplasmic reticulum with corresponding enzymatic induction phenomena. The latter occur in the hepatic microsomes and are obviously associated with the detoxication reactions of the contaminated organism. Proliferation of the endoplasmic reticulum is one of the first signs of physiotoxicological action in response to exposure to micropollutants that we can rely on at the present time.

The associated enzymatic induction was discovered by chance by the pharmacologists Hart *et al.* (1963) who were studying the effects of a barbiturate, hexobarbitone, on rats. These researchers were surprised to find that a dose of hexobarbitone which previously induced eight complete hours of sleep in their animals, was suddenly ineffective. Detailed checks led them to identify the one element that could have modified the physiological conditions of their experimental animals: the spraying in the animal cages of an insecticide, chlordane, which had been done in order to rid the rats of their external parasites. This organochlorine compound had, therefore, shown itself to be antagonistic to the drug being studied, hexobarbitone. The scientists could then prove that chlordane exercised this effect by accelerating the localized synthesis of enzymes in the microsomal fraction of the hepatic cells. The hydrolases and oxidases thus produced in this intracellular fraction break down the barbiturates into inactive, water-soluble by-products, which are eliminated by urinary secretion. It is now known that this reaction of hepatocytes to toxic substances is fairly general. Experiments have shown that in insects the microsomes of the adipocytes play a similar role.

The processes of enzymatic induction are also caused by carcinogenic compounds. From 1956, Conney *et al.* demonstrated the increase in the synthesis of demethylase in the hours following administration orally or into the abdomen of 3-methylcholanthrene in concentrations varying from 0.5 to 50 ppm. These authors could also show that this extremely carcinogenic compound stimulated at the same time the synthesis of other enzymes associated with the hepatic microsomes, from the group of reductases.

(b) Renal lesions—Many toxic agents cause renal lesions, particularly in the syndromes resulting from acute exposure. However, what is more worrying is the induction of renal necrosis by various substances after long-term exposure. There are countless examples of these effects, one of which results from taking

persistent doses of phenacetin, the aspirin substitute, whose long-term usage leads to irreversible renal necrosis.

Mercury poisoning is another well-known cause of serious nephritis. Various pesticides have similar effects. For instance, morphamquat, a herbicide from the bipyridilium group, induces changes in the Malpighian glomeruli and the collecting tubules of the kidneys after both acute and long-term exposure to it, (Ferguson et al., 1969). Likewise, diphenyl, a fungicide used to protect citrus fruits from attacks by various phytopathogenic agents during transport, has also been shown to be a necrotic agent to the kidney.

2. Principal germinal effects

The action of chemical or physical biocides can reduce the numbers of a contaminated population, not by causing the death of the polluted individuals but by affecting their biotic potential, either by direct sterilization of the adults or by causing teratological changes which reduce the viability or fertility of the offspring from contaminated parents, or even by the induction of lethal mutations in the ova or uterus and sublethal mutations which decrease the number of offspring produced.

Numerous toxic agents, both physical and chemical, can sterilize living beings. The effect of ionizing radiation is well known in this context. It would happen as a result of a greater sensitivity of the gametes, the active reproductive cells, to the rays.

Effects of the Biotic Potential

A whole series of chemical compounds possess sterilizing properties. Among them are numerous pesticides: the effects of organochlorine insecticides on the reproductive functions of birds was first demonstrated in 1956 by Genelly and Rudd. Since then, it is now understood that these effects come about as a result of the neurotropic action of these substances. They induce neuroendocrine damage which affects the process of sexual maturity. Furthermore, their toxic effects also cause an acceleration of the metabolism of oestrogen hormones, which results in irregularities in the hormone system. Kupfer (1969) showed for example that 25 ppm of DDT given to rats accelerates the degradation of all steroids and in particular doubles the speed of catabolism of testosterone, 17-β- oestradiol and progesterone.

Many other micropollutants affect the reproductive functions of vertebrates. One of the most disturbing problems is the presence of polychlorinated biphenyls (PCBs) in the environment, which are substances used extensively in the production of plastics. PCBs cause various malfunctions in the ovaries of birds, in the same way as many organochlorine insecticides. The direct results of these malfunctions are thinning of the eggshell and its correspondingly greater fragility (cf., for example, Hickey and Anderson, 1968).

Mutagenic Properties

Other pollutants can diminish the biotic potential of animals by direct germinal effects. Ionizing radiation is one of the better-known mutagenic agents. Some lethal mutations following exposure to radiation can affect the gametes and modify their genome so that although their activity is not impaired, they produce non-viable zygotes which are incapable of forming the embryo correctly.

There are also chemical substances called radiomimetic because they act in a similar way, and can produce the same mutations. Examples of these are the chemosterilants of the aziridine group: apholate, tepa, metepa.

Teratogenesis of the Gonads

Apart from the classic teratogenic effects, there are other more damaging consequences caused by various natural and synthetic compounds which can sterilize either in the egg or in the uterus the offspring of contaminated individuals. These teratogenic effects occur, therefore, at the embryonic gonad stage.

Lutz-Ostertag and Lutz (1969, 1973) were able to show that various insecticides (aldrin, DDT, parathion) and herbicides (2, 4-D, simazine, 2, 4, 5-T) induced numerous abnormalities in the development of the sexual glands of both sexes in the embryos of quails, chickens and pheasants. These were characterized by malformations of the reproductive system, giving rise to bisexual offspring possessing Müller canals that were either undeveloped or intact. The researchers also noticed that the embryos had atrophy of the ovarian cortex, with no Pflüger's cords, and therefore no gonocytes. The *in vitro* study of the action of these pesticides on organotypic cultures of embryonic gonads confirms the results obtained for the whole egg (Didier, 1974, 1981).

B—PATHOLOGICAL PROBLEMS PECULIAR TO ECOTOXICOLOGY

1. Permanent exposure

Research into the mechanisms and effects of chronic toxicity and poisoning, in spite of their physiological interest, are generally only of limited usefulness in ecotoxicology which for the most part involves examining the effects of *permanent* exposure to very low concentrations of substances present in the environment. Another difficulty from the point of view of methodology and experiment is the fact that man's environment and numerous ecosystems are today being increasingly contaminated by a multitude of micropollutants belonging to a large variety of chemical groups. These conditions are a long way from those of the laboratory where the toxicologist usually works on pure substances taken in isolation. In fact, ecotoxicology's most specific field of study is that of the pathological consequences for living beings exposed simultaneously to micro-doses of the different toxic substances polluting the environment.

The presence of synergistic and antagonistic factors makes these ecotoxicological problems particularly worrying at the present time. Synergism is a well-known factor which has been used profitably in the past as a means of enhancing the strength of pesticides. The effectiveness of pyrethrin insecticides can in fact be increased tenfold if they are stimulated by an equal amount of piperonyl butoxide, a substance which used in isolation is only an extremely weak insecticide. The synergistic factor can, unfortunately, also occur accidentally when a toxic compound interacts with a natural substance or with a micropollutant present in the air, in water or in foodstuffs. One spectacular example of this happened recently when it was discovered that serious cerebral vascular lesions occurred in individuals who had eaten fermented cheeses, rich in tyramine, while on the antidepressant drug tranylcypromine. The drug is a strong inhibitor of monoamine oxidase and, therefore, blocks the degradation of the ingested tyramine.

2. Consequences

Some of the most serious pathological risks associated with prolonged exposure to micropollutants in the environment are those whose effects are characterized, not by specific action on a particular physiological function, but rather by their ubiquity, and can have dangerous long-term consequences for the affected individuals or their offspring. A few of the reactions caused, particularly allergies and carcinogenesis, are disorders of a somatic nature, and others such as mutagenesis and teratogenesis produce both somatic and germinal effects.

Allergies

These provoke varied clinical responses: asthma, eczema, visceral or vascular disturbances, mechanisms producing immune responses and individual sensitivity due to certain genotypic characteristics. Reactions are usually to a group of chemicals rather than to pure substances (Gervais, 1976).

Mutagenesis

This results from the action of various chemical or radioactive micropollutants. The incidence of mutations is one of the major preoccupations of ecotoxicology in an age when the use of certain substances and/or the exposure of the whole of humanity to ionizing radiation are posing so many potential threats to our genetic heritage.

Furthermore, it is now known that mutagenesis is the first step towards carcinogenesis on the somatic level. Current fears about the dispersion of polyvinyl chlorides or the proliferation of nuclear power-stations illustrate the

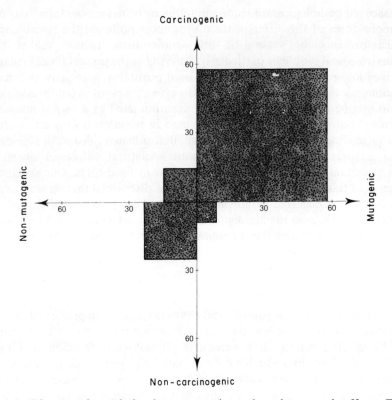

Fig. 1.7 Diagram of correlation between carcinogenic and mutagenic effects. There appears to be a positive correlation for over 80% of the substances tested which display mutagenic or carcinogenic properties. (After Sarasin, 1975 in *La Recherche*. See under Ames, 1975.)

topicality of these questions. Basically, mutagenic effects alter the genetic code through the interaction of certain chemicals or of ionizing radiation with the base sequences of DNA, that is with its polydeoxyribosephosphate skeleton. The changes that follow affect some codons or can be even more drastic and result in a break in the chain. When this occurs, the damage is visible on the cytological level and results in the breaking of chromatids, the translocation of chromosomes, the creation of supernumerary chromosomes, etc. If the damage affects the reproductive cells, irreversible abnormalities will occur in the genome. If somatic cells are affected, it can be the cause of a carcinogenic process.

Some recent research has in fact shown that there is a high probability of correlation between mutagenesis and carcinogenic potential. MacCann *et al.* (1975), using the Ames test (see below), have studied the mutagenic influence of some 300 different compounds belonging either to chemical groups known to be carcinogenic or to chemical groups not linked with these pathological effects: aromatic amines, various organohalogen derivatives, aromatic and heterocyclic hydrocarbons and their nitrated derivatives, nitrosamines, diazo compounds, aflatoxins, carbamates and various heterocyclic compounds. They were able to show that 90% of carcinogenic substances are mutagenic and that, conversely, none of the substances tested which were known to be non-carcinogenic proved themselves to be mutagenic. Only asbestos and certain hormones are carcinogenic and non-mutagenic. Examples of the rare non-carcinogenic mutagens are bromouridine and hydroxylamine.

(a) Mutagenic tests—Perfection of the Ames *et al.* test (1973) enables very rapid detection of the mutagenic potential of a substance. The test determines whether the substance exercises a mutagenic effect on a mutant histidine (his-) of the *Salmonella typhimurium* bacteria. The addition of a mutagenic substance to such a his − bacteria will cause the production of reverse his − his + mutants

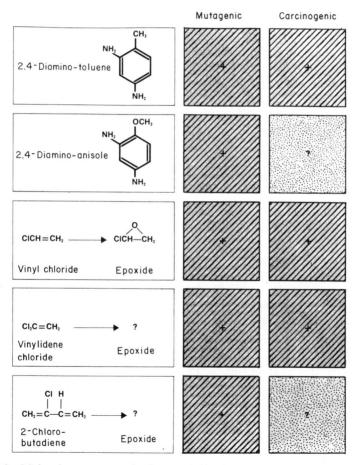

Fig. 1.8 Molecular structure of a few synthetic chemicals proved to be mutagenic by the Ames test. The mutagenic and carcinogenic activity is indicated by a cross. (After Sarasin, 1975 in *La Recherche*. See under Ames, 1975.)

capable of synthesizing the histidine and, therefore, of growing in a culture medium without histidine, as opposed to the spores of the initial strain. Each time that a substance tested presents mutagenic properties, colonies of *Salmonella* will be obtained in a medium without histidine.

The interesting feature of this particular type of bacteria is that the sequence of its histidine gene is known exactly, which means that the DNA sequence attacked by the carcinogen being tested can be determined. For example, it has been possible to show that certain polycyclic hydrocarbons react at the level of the base sequences C-G-C-G-C-G-C-G of the histidine gene!

This flexible method can, for example, by adding some microsomal enzymes to the culture medium, be used to determine precisely the mutagenic and carcinogenic potential of the degraded products of a toxic substance. It is a fact that an apparently non-carcinogenic compound can be transformed inside the organism into carcinogenic substances. The preliminary incubation of the chemical being tested with microsomal samples from the liver, kidney or other organs means that the mutagenic potential of these metabolites can be determined. This variation of the Ames test (cf. Fig. 1.8) has led to confirmation of the serious evidence concerning the carcinogenic potential of vinyl chloride and 2-chlorobutadiene. It has also shown that 89% of hair dyes are mutagenic and added more proof of the carcinogenic properties of cigarette smoke.

According to MacCann *et al.* (1975), the lack of mutagenic potential in 10% of the carcinogenic compounds they tested could be due either to incorrect metabolic activation of the substances *in vitro* linked to inadequacies in the technique they used with those particular substances, or to the fact that those particular compounds have doubtful carcinogenic potential (as in the case of pararosanilinene).

The micronucleus test, developed more recently by Schmid (1975), constitutes another method of detecting mutagens based on the presence of Howell–Jolly bodies in the young erythrocytes in the bone marrow of rats. The interesting feature of this technique lies in its ability to show direct proof of a mutagenic effect on mammals. The mutagenic agents cause chromosomal breakages during the mitoses of the erythrocyte system. In this way, residual nuclear remnants, called *micronuclei*, can be observed in the polychromatophid erythrocytes which expel their nucleus a few hours after the end of the last mitosis. The experimental procedure of the test consists of administering two doses of the chemical being studied at an interval of 24 hours, followed by the dissection of the infected rats 6 hours after the second dose. The bone marrow is extracted and used to make coloured smears following the Pappenheim–Unna technique, so that the eventual micronuclei can be seen in the erythrocytes. The application of this method has enabled Siou *et al.* (1977) to show the strong mutagenic potential of benzene and benzopyrene.

Carcinogenesis

Although there is as yet no complete explanation for the mechanisms of the processes which govern it, carcinogenesis appears on the somatic level as a result of changes made in the genetic code of a particular cell in the organism. It is now becoming clear that the damage to DNA caused by carcinogens plays a fundamental part in the initiation of cancer. Carcinogenesis is, therefore, the product of a disturbance in the mechanisms which control the repression of certain genes in the differentiated cells of the tissues of multicellular plants and metazoa. The result is that the affected cells lose their specialization (dedifferentiation) and begin to proliferate uncontrollably.

Cancers are a group of disorders affecting all human races and ages. They are also found in all animal species, both vertebrates and invertebrates. There are three broad categories of cancers as defined by histopathological criteria: sarcomas affect connective tissues, carcinomas affect epithelial tissues and lymphomas affect blood tissues.

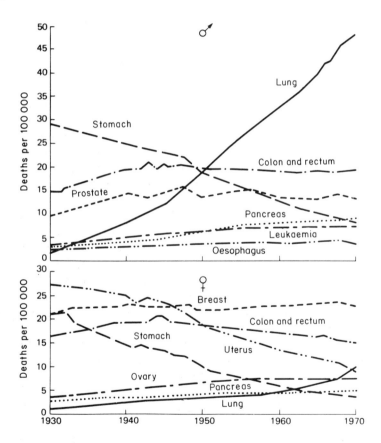

Fig. 1.9 Variations in the mortality rate due to cancer in the United States between 1930 and 1970 according to sex and the affected organ (adjusted to the population of 1940). Not the considerable rise in lung cancer linked to the increase in the smoking habit. (Source: David L. Levin et al., Cancer Rates and risks, DHEW Pub. No. (MH) 75-691 (2nd ed: Washington, DC: Government Printing Office, 1974), p. 14.)

All available biomedical statistics indicate a serious rise in the general frequency of tumours since the beginning of this century in industrialized countries. For example, in the United States, their incidence has doubled since 1935. Also in the USA, tumours were the cause of 3.7% of the mortality rate in 1900 and 15.6% in 1970. The change in these figures has in fact been very different according to the type of cancer: tumours of the pancreas and lung have increased considerably over this period (there are about twenty times as many cases of lung cancer in keeping with the appalling rise in tobacco smoking), as opposed to stomach cancer, which was the primary cause of cancer mortality at the beginning of this century and has diminished notably for reasons not yet fully understood (cf. Fig. 1.9).

The Role of Environmental Factors

There are now many epidemiological studies which prove the major role played by environmental factors in the induction of cancer tumours. Allowing for the fact that some cases are unquestionably endogenous, due to genetic deficiencies (*Xeroderma pigmentosum*, for example, although even here, the initiation of the disease originates in exogenous factors), it has been estimated that more than 80% of tumours are caused by exogenous factors related to life-style and especially to environmental pollution. The correlation between smoking habits and cancer of the lungs or larynx is most striking. Examples of other factors are: permanent exposure to numerous carcinogenic chemicals at the place of work or even in the home and exposure to physical carcinogenic agents; excessive sun-bathing (ultraviolet) or irradiation from ionizing radiation.

Man's environment in the technological civilization is quite simply saturated with an enormous number of substances that have carcinogenic properties. Some of them, such as asbestos or vinyl chloride, were considered to be inoffensive and were, therefore, used for decades on a vast scale before the danger they posed was recognized. It is also evident that the large increase in synthetic organic substances — several thousand new compounds are commercialized every year — poses difficult and urgent problems when it comes to evaluating their toxicological properties.

Since the discovery in 1775 of cancer of the scrotum in chimney-sweeps, epidemiological studies of certain groups of workers in the chemical industries, as well as animal experiments, have shown that a large number of mineral or organic compounds — either synthetic or of natural origin — have carcinogenic properties. During the 1930s, the abnormal frequency of cases of cancer of the bladder in workers involved in the production of benzidine, an essential raw material for the dye industry, gave irrefutable proof of the role of environmental pollution in the genesis of tumours. Whereas cancer of the bladder accounts for 13.2 cases per 100 000 cancer cases in the population of America, it reaches a level of 21% in average amongst people who have worked in benzidine production.

The first cases of pulmonary mesothelioma, an otherwise extremely rare disease, associated with professional exposure to asbestos, were noticed in 1950.

Table 1.1 Latency period preceding the onset of tumours in workers exposed to 78 aromatic amines. (After Wilhelm Hueper, Medicolegal Considerations of Occupational and Nonoccupational Environmental Cancers, Ch. VII in Charles Frankel, ed., *Lawyers' Cyclopedia: and the Land*)

Latency period (in years)	Percentage of workers with tumours as a function of the length of their exposure to the amines (in years)					
	up to 1	1	2	3	4	5 and over
up to 5	0	0	0	0	0	0
10	0	0	0	0	0	11
15	0	17	22	0	10	45
20	4	17	22	40	30	69
25	9	17	22	70	70	88
30	9	17	48	70	80	94

Since then, the rapid increase in asbestosis is obviously linked to the considerable rise in the use of asbestos as a thermal insulator, building material and in the manufacture of various components for motor vehicles.

It was only in 1974 that the role played by vinyl chloride in the induction of hepatic angiosarcoma in workers in the plastics industry was discovered. Subsequent research has revealed that exposure to this substance also causes an increase in cancer of the respiratory and lymphatic systems.

At the present time, the atmosphere, water and human foodstuffs in the industrialized countries are contaminated by a very large number of substances that have carcinogenic properties. One of the major difficulties in identifying the potential toxicity of a chemical compound, apart from the enormous numbers that require testing, lies in the length of the latency period from the time of initial exposure to a carcinogen to the appearance of the clinical symptoms of cancer (cf. Table 1.1). It is acknowledged that a few decades can elapse before pathological signs appear. Thus, the supposedly harmless use of a substance for ten or fifteen years can give it a reputation for being safe when in fact it can cause a high frequency of cancer after a latency period of about thirty years. In the case of aromatic amines (cf. Table 1.1) it can be seen that tumours may occur thirty years after only a few months' exposure to an environment contaminated with these substances.

In any case, epidemiological studies cannot constitute an acceptable procedure for controlling the potential danger of chemicals because, according to this hypothesis, it must be accepted *de facto* that a group of individuals should act to a certain extent as human guinea-pigs. In the most favourable circumstances, the results of these studies lead to banning *a posteriori* the use of chemicals which have already, and sometimes for years, created havoc with public health.

Experiments with animals would appear, therefore, to be indispensable for detecting potential carcinogens. They can, however, have the drawback of taking a long time to do—months and even years. A further difficulty is the immense quantity of chemicals present in man's environment, numbering more than 2 million different compounds. Only some 6000 of them have been tested so far in the laboratory for their carcinogenic properties, although specialists estimate that between 10% and 15% of all organic compounds may be carcinogens.

The principal known carcinogens are classified as in Table 1.2. It appears in fact that carcinogenic substances are not distributed in an erratic way but correspond to well-defined groups of chemical compounds.

There are large variations within each category depending on their action and precise structure. For instance, bis-chloromethylether is a potential carcinogen in concentrations to the order of a few parts per billion. In the same group of compounds, bis-chloroethylether appears to be a much weaker carcinogen

Table 1.2 Principal classes of carcinogens

Class or example	Principal cause of exposure for man
Ultra-violet radiation	Excessive exposure to sunlight (sunbathing)
Ionizing radiation	Medical, scientific and industrial use of X-rays and γ rays
Radionuclides	Medical and scientific uses and in the nuclear industry
Aflatoxins	Natural substances arising from the culture of certain fungi in food products
Steroid hormones	Food additives for domestic animals
(diethylstilboestrol)	Drug
Polynuclear hydrocarbons	
Anthracene	Tobacco smoke
Benzopyrene	Combustion of coal, fuel oils and petrols
Organohalogen compounds	
Vinyl chloride	Monomer used in the plastics industry and with a trace occurring in foodstuffs wrapped in plastic (PVC)
Aldrin—Dieldrin	Insecticides
Bischloromethylether	Impurity in certain synthetic resins
Aromatic amines:	
—β-naphthylamine	Synthetic rubber
—Benzidine	Production of dyes
—Diazoics	Food colouring (no longer used) for making
p-dimethylaminoazobenzene	margarine yellow (butter yellow)
—Cyclamates	Sweeteners, food additives
—Nitrosamines	Rubber additives. Apparently produced by the metabolism of nitrites contained in foodstuffs
Metallic dusts:	
—Beryllium	Light alloys
—Chromates	Paint pigments
—Cadmium	Electronics industry, stabilizing plastics, etc.
—Arsenic	Pesticides, pharmaceutical industry

needing concentrations a hundred times stronger in order to induce the same effects.

To conclude, there are a few things to be said about the validity of animal experiments. For obvious material reasons, it is not possible to work in a laboratory on large numbers of animals. Supposing that a compound in a given concentration induces a carcinogenic effect in one in a thousand individuals: a histopathological test carried out in the laboratory on a hundred rats would have a strong chance of not observing the carcinogenic effect. However, an incidence of 1°/00 (one in a thousand) in terms of human populations for cancers caused by the use of such a chemical would be absolutely unacceptable. Taking account of experimental limitations, it then becomes necessary to work with much stronger concentrations of the chemical than those encountered in man's environment and to work on a restricted number of experimental animals, about 50 to 300 per dose as a general rule. This implies, therefore, a knowledge of the dose-effect relationship in order to extrapolate results for weaker concentrations, as will be discussed in the next section (see page 33).

Finally, as most experiments are carried out on rodents, a certain margin of error is involved when results are extrapolated to apply to man, bearing in mind differences in specific sensitivity. This goes beyond the subject of carcinogens and constitutes a general problem in ecotoxicology.

Teratogenesis

This affects the offspring of individuals exposed to various pollutants. From it can result either abnormalities in the development of the embryo in the egg (*in ovo*) or in the uterus (*in utero*), due to exposure to toxic agents, or it can induce non-lethal mutations in the genotype of the offspring before fertilization.

Various pollutants, such as pesticides, display strong teratogenic potential. The large-scale use of a defoliant, 2,4,5-T, during the Vietnam conflict provoked legitimate protests at the end of the 1960s following the discovery of congenital abnormalities in the children of women who had been living in the densely treated zones. Subsequent studies have shown that the major part, if not all, of the observed teratogenic effects were due to an impurity, dioxin, which hit the headlines yet again with the Seveso catastrophe in July 1976 (discussed later).

The case of thalidomide, the hypnotic drug responsible for serious abnormalities in the embryogenesis of the limbs of the embryos exposed to this substance between the 23rd and 40th days of pregnancy, also illustrates well the dangerous consequences that can result from the absorption of teratogenic agents.

C—THE DOSE-RESPONSE RELATIONSHIP IN ECOTOXICOLOGY

A particularly fundamental question which permanently confronts ecotoxicology is how to determine the exact relationship between dose and effect for long-term exposure to the various micropollutants which contaminate the biosphere.

1. Dose accumulation and genotoxic effects

It is important to know if the continual absorption of toxic substances in infinitesimal amounts can or cannot induce genotoxic effects (mutagenesis, carcinogenesis). A topical example, for which no definitive answer has as yet been given, is how to determine the rates of mutagenesis and carcinogenesis in human populations which will result from the slow but inevitable growth of radioactivity in the natural environment connected with the nuclear industry.

The Dose-Response Curve

In other words, we need to know:

(a) whether the accumulation of effects caused by exposure to traces of micropollutants is total, partial or nil; and
(b) whether there is a threshold below which no effect will result from the action of a toxic chemical or radiation, or whether on the contrary, any exposure, however weak, has minimal but irreversible consequences.

The answer to this type of question is extremely complex. In many cases, we do not yet know what the dose–response curve really looks like for long-term exposure to low concentrations of micropollutants, in spite of its crucial importance to the protection of our environment.

There are four distinct categories of dose–response curves corresponding to the following possibilities:

1. The long-term effects of permanent exposure to low doses are more dangerous than taking an equivalent dose of the toxicant in a short period of time.
2. There is a summation of irreversible effects whatever the duration and means of contamination.
3. The accumulation of doses absorbed per unit of time is only partial because of efficient detoxication and repair mechanisms, and there is no absolute summation of irreversible effects.
4. The effects of low doses are beneficial, with toxicity only appearing after certain concentrations.

The dose–response curves corresponding to the first two possibilities have no threshold, whereas the second two imply that a threshold exists.

There are many concrete examples which demonstrate the large variety in the appearance of dose–response curves according to the nature of the substance and to the toxicological effect being studied.

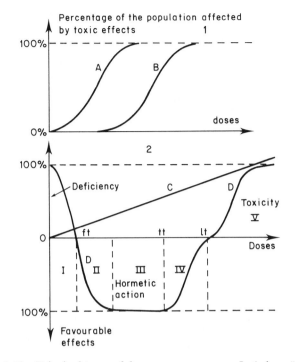

Fig. 1.10 Principal types of dose-response curves. In 1 sigmoïds are shown without a threshold (A) or with a threshold (B).

In 2, the straight line C represents the possibility that the dose-response relation is linear, thus implying the absence of any threshold. In D, a more complex possibility is shown where a substance which has beneficial effects in low doses becomes toxic beyond a certain dose level. (ft = threshold from which favourable effects are apparent; tt = threshold of toxicity; lt = lethal threshold)

Examples

Various carcinogenic compounds have dose-response curves of the first type defined above, which makes them particularly dangerous. In an essay which is now a classic, Druckrey and Kupfmuller (1949) *in* Truhaut (1956) proved the cumulative effects of a powerful carcinogenic agent, paradimethylaminoazobenzene, better known as butter yellow (a colouring formerly used in the production of margarine). They demonstrated the strictly cumulative action of this chemical in the induction of hepatomas in rats, which was in spite of the importance of degradation and excretion mechanisms. Subsequently, Druckrey *et al. in* Szakvary (1965) showed that paradimethylaminostilbene caused cancer of the auditory duct in rats when administered in cumulative doses that were weakened as the exposure time was increased (cf. Table 1.3).

Carcinogenesis is caused by an exposure time of 250 days to a dose of 3.4 mg/kg/day or a cumulative dose of 850 mg. However, a cumulative dose

Table 1.3 Dose-effect relationship in the induction of cancer of the auditory duct in rats by paradimethylaminostilbene (Druckrey et al., in Szakvary, 1965.)

Daily dose (in mg/kg)	Time to occurrence of effect (in days)	Total dose needed to cause carcinogensis effect (in mg/kg)
3.4	250	852
2.0	342	685
1.0	407	407
0.5	560	280
0.2	675	135
0.1	900	90

of only 90 mg is needed if exposure time is increased to 900 days with a dose rate of 0.1 mg/kg/day.

Types of Dose-Response Curves

(a) Straight-line graphs—The dose-response curve seems to be linear under certain circumstances, even with low concentrations. This appears to be the case with some of the effects of ionizing radiation. The report published on that subject by the National Academy of Sciences of the USA (1972) admits that the occurrence of myeloid leukaemia corresponds to a 100% cumulative effect, with the incidence of this disease rising to 3 cases in 10^6 (1 million) people per year and per rem. Linear dose-response curves are also more probable in cases of the onset of mutagenic effects in mammalian cells in a culture medium exposed to low concentrations of aflatoxin or Captan (cf. Legator *et al.*, 1969).

It may not yet be possible to draw any categorical conclusions with both of the latter examples, but it has to be admitted that the dose-response curves would have no threshold. Such an admission is particularly serious in the context of present ecotoxicological problems like the pollution of human populations by various organohalogen compounds (DDT, chlorinated cyclodienes, PCBs). Consider what risks must be associated with irradiation, affecting the lungs of the world populations, from the ^{85}Kr (krypton) being emitted into the atmosphere by light-water nuclear reactors whose number would double every three years until the year 2000 if certain technocratic plans are followed.

(b) Sigmoid curves—The second type of dose-response curve results from the existence of detoxication and repair mechanisms which intervene at the cellular level or in the organs and are all the more efficient when chronic exposure is being caused by weak doses. To this type belong a whole series of reactions of a population to a toxicant with a range of harmful effects descending in importance from lethal responses down to biochemical effects which can be detected analytically but which have no associated pathological actions (cf. Fig. 1.11). In this instance, the dose-response curves will take on the

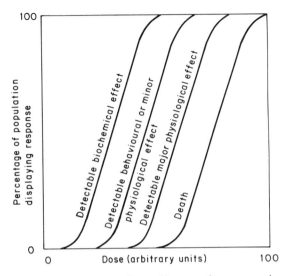

Fig. 1.11 Influence of a gradient of increased concentration on the physiotoxicological response of an organism. When dose levels are very low only slight biochemical modifications are detectable which cannot be associated with any pathological change

appearance of a sigmoid with a threshold that enables the definition of a minimum threshold of exposure from which harmful effects can be detected and an area of dose without effect. For example, it seems that the mutagenic effect of ultra-violet radiation is of this type.

The existence of detoxication processes is also a factor that helps to diminish the relative toxicity of weak doses: ethyl alcohol, which is metabolized fairly easily by mammals, seems to have a toxic threshold. The same applies to various other toxic substances. In our own researches we were unable to show up any significant effects on the lifespan of insects by insecticides applied in sublethal doses.

(c) Graphs showing antagonistic effects—The third type of dose–response curve, and more complex than the other two, corresponds to the situation where natural substances necessary for organisms in very small doses can become dangerous and even extremely toxic in greater concentrations. A whole series of mineral and organic trace elements present such curves, for example cobalt, fluoride, vitamins A and D, etc. They are all fundamental constituents of the living cell or of certain organisms, indispensable as a slight trace in food but if absorbed continuously, even in relatively low doses, they can cause serious disorders, such as fluorosis, hypervitaminosis A, etc.

Fluorine, for instance, is an essential element for ossification, but man cannot absorb permanently more than 1.5 ppm of it in drinking water without experiencing side-effects. An excess of fluorine causes a serious illness, fluorosis, whose symptoms are problems first of all with teeth and bones, developing into progressive cachexia. Persistent intake of fluorine in food has the strange effect of causing serious disorders even when absorbed in a far lower concentration than that needed to cause acute poisoning from a single dose. Just a few centigrams per day in repeated doses are sufficient, whereas a whole gram of NaF is needed to cause acute poisoning.

Luckey and Venugopal (1977) proposed the term 'hormetines' to designate any substance (either an element or an organic compound) with properties which in low doses are stimulating for the physiology of organisms but which become toxic above a certain concentration. As well as the mineral or organic trace elements already mentioned, the authors estimate that perhaps some 50% of all potentially toxic substances would have a beneficial effect on living things if taken in low doses. For example, it is well known that various insecticides increase the fecundity and fertility of female insects in sublethal doses (cf. Ramade, 1967). With such substances, the dose–response curve would of course have not one, but two thresholds. In Fig. 1.10 curve D shows the variations in the response of an organism to increasing doses of a trace element (cobalt, for example). The areas I and II on the graph represent insufficient concentrations (deficiency zone). In area I, the deficiency causes a mortality rate that decreases with the concentration. In area II, the ft (favourable threshold) is reached, from which favourable effects begin to appear. However, area II also represents a deficiency zone as the concentrations of the substance being studied are insufficient to sustain a normal physiology. Also, essential biological processes such as growth or reproduction will take place even below a normal level. In area III we reach the zone where the concentrations of the substance are those met under usual ecological conditions. This is the area where the optimum favourable physiological effects are encountered (the zone of hormetic action). In area IV we reach the tt (toxicity threshold) beyond which appear the toxic effects associated with an excessive concentration of the substance. In area IV the toxic effects will cause physiological deficiencies such as slower growth. Finally, in area V we enter the lethal zone and find a classic dose–response curve, the sigmoid type with a threshold.

Cumulative potential

The existence of a potential for cumulative action of absorbed doses of numerous toxic substances is a very worrying problem. Many pesticides, for example, display this potential. It is shown by the appearance of physiotoxicological disorders after the intake of a total quantity of the toxic chemical which is considerably less than the single amount needed to cause acute poisoning. An example of this is that the LD_{50}, or median lethal dose of β-BHC taken orally in a single dose is one of the highest for rats, namely about 6000 mg/kg. In contrast, however, if the same animals are fed consistently with doses of less than 10 mg/kg, this will cause serious hepatic lesions within a few months.

Another worrying aspect of the potential for cumulative effects of certain micropollutants to which man can be permanently exposed, is the fact that they may be transferred to the offspring. For instance, if pregnant rats are given food contaminated by strong carcinogens (nitrosamines) the result is that their offspring develop brain tumours when they reach maturity, in spite of there being apparently no ill effects to the pregnant rats themselves from the drugs they were given (a case of transplacental carcinogenesis). The transmission of cumulative effects has also been observed after cases of pollution by heavy metals: in Minamata, for example, a variety of congenital defects, due to accumulation of methyl mercury in the foetal brain, affected children who had been exposed to the substance *in utero*.

2. The notion of the maximum admittable concentration

One of ecotoxicology's most essential tasks is the study of the harmful effects on warm-blooded vertebrates of permanent exposure to the principal pollutants, and in particular, a knowledge of the dose–response curves which characterize those pollutants. It is also of prime importance for the protection of human health in our modern technological civilization, due to the ever-increasing contamination of our environment by innumerable toxic substances. Air and food, the two things on which man depends completely from birth, are also the most vulnerable to pollution and have, therefore, been the subject of a large number of ecotoxicological studies. These investigations have led experts to define the maximum so-called 'acceptable' dose and the corresponding admittable concentration (MAC) for the principal pollutants in our environment. The calculated doses are considered to be harmless for our species, even after life-long uninterrupted exposure. In the same way, maximum tolerable levels have been established for noise (in the place of work) and for ionizing radiation (for workers in the nuclear industry and for the rest of the population).

Evolution of the Notion of an 'Acceptable' Dose

A retrospective examination of the norms laid down over the last decades shows that the various committees of specialists charged with fixing maximum 'acceptable' doses have in fact tended to revise the levels set by lowering them progressively as the physiotoxicological knowledge of each pollutant increased. The example of DDT is very significant. The experts of the FDA (Food and Drugs Administration in the United States, responsible for legislation and monitoring of foodstuffs, medicines and cosmetics) finally recommended a zero level of residual DDT insecticide to be permissible in milk, before going on to prohibit its further use over the whole of the USA. In the case of mercury, a growing number of toxicologists consider that the maximum level of it allowed by the WHO in human foodstuffs, 0.5 ppm, is too high and should be reduced at least to 0.2 ppm.

There is no general answer to the problem of fixing maximum 'acceptable' doses. Everything depends on the type of dose–response curve characterizing the pollutant in question, as was shown in the examples given earlier.

Examples

It is certainly not advisable to take as a general rule the few well-known examples of substances displaying a distinct threshold below which no harmful effects exist. When long-term exposure can produce positive accumulation of the doses absorbed, it must be acknowledged that the dose–response curve has no threshold. At the present time, it appears that the majority of the mutagenic and carcinogenic effects of ionizing radiation and of most chemical compounds with similar properties have this type of dose–response curve. In this eventuality, it must be agreed that it is no longer possible to talk about a maximum 'acceptable' dose nor of a maximum dose without effect.

The Threshold Limit Value

All of these considerations have led toxicologists to substitute the notion of a 'threshold limit value' for that of a 'permissible dose' and to use the concept of the benefit/risk relationship to justify the thresholds laid down.

Most human activities imply a risk of accident and, therefore, of death, whether they involve the deliberate decision of an individual or involuntary exposure to a natural danger (flood, disease) or to an artificial danger (technological developments). Table 1.4 shows the mortality risks associated with various voluntary or involuntary activities, with accidents arising from natural or artificial causes and with diseases.

How else then can the use of certain substances whose carcinogenic properties are well known be justified? At the congress on carcinology held in Florence in October 1974, the example was given of 17 potentially carcinogenic substances

Table 1.4 The probable mortality rate associated with various risks of both technological and natural origin. (After Holdgate, 1979 and Upton, 1982.)

Type of risk	Associated mortality rate (number of deaths/ 10^5 people/year)
Voluntary activities[1]	
Smoking (20 cigarettes/day)	500
Alcohol (1 litre of wine/day)	75
Driving a car (GB)	14
Riding a motorbike (USA)	2000
Diseases, natural disasters, pollution	
Influenza	20
Leukaemia	8
Walking on the roads (GB)	6
Pollution from nuclear power stations (USA)	0.25
Floods (USA)	0.22
Earthquakes (California)	0.25
Transport of inflammable products	0.005

[1] These data are calculated on 100 000 people being exposed to the risk, and do not represent the actual number of deaths recorded in a population of 100 000.

of which 9 are produced, all with good reasons, in quantities of at least 1000 tonnes per annum world-wide. The term 'safety threshold' was banned during the discussions concerning maximum doses of these chemicals for both the professional sphere and for the world's population, because such substances cannot be considered as having no effect, even when they do not noticeably reduce life-expectancy. In reality, the micro-doses to which inhabitants of industrial zones or workers in chemical industries are exposed, do present cumulative effects whose consequences can sometimes be felt after more than twenty or thirty years of exposure to them.

Who Takes the Risk?
This question raises one of the most worrying aspects of modern technocratic ideology. The taking of risks (generally on behalf of others) in the field of toxicology is in fact the business of committees of experts drawn up by governments and among whom scientists—who are the only people really competent on the ecotoxicological level—are most often in the minority. It is these committees, usually made up of high-ranking civil servants and representatives of the industries concerned, who are responsible for creating the balance between possible risks to human health and eventual benefits that can result from using a particular substance.

Under such conditions it can be seen that many decisions whose long-term repercussions could be extremely important for the fate of whole populations, are being taken by administrative bodies of questionable competence, given the multitude of new problems which can arise and are recognized only by a few experts. Furthermore, in the industrialized nations, the decision to promote a particular chemical or to develop a new technology (as is the case with nuclear power, for example), has always been taken in the past before evaluating its exact ecological consequences.

It was not just a coincidence that certain toxicologists—notably Truhaut and De Tomatis—at the 1974 Florence congress on carcinology deplored the fact that it had taken the appearance of cases of hepatic angiosarcoma among workers in the production of vinyl polychlorides before difficult and costly measures were eventually taken. It is also worth recalling the fact that some years ago animal experiments with vinyl chloride, although insufficient, had proved the carcinogenicity of this product, but that these results were not taken into account by the committees responsible for controlling its production.

In the same way, it is not just chance that from 1970 onwards the US National Academy of Sciences asked the United States government to give 'to an organization truly independent of the body responsible for radioactive pollution (the AEC) responsibility for the control of all radioactive matter released into the environment'.

Missing Information
One of the least trustworthy and most disturbing aspects of the benefit/risk concept is that each committee of experts involved tends to define in a rather

one-sided way a maximum tolerable dose for each pollutant *in isolation* or at most for a mixture of a few substances of the same group (pesticides, for example). This is a serious deficiency because it means that on the epidemiological level the possibility could be ignored of potentiation between toxicants with different applications (for instance, the effects of pesticides on food additives). Above all, it forgets to consider the complete range of micro-doses to which man is exposed.

All this is even more surprising given that the literature of toxicology actually contains an abundance of examples of synergism. They are found in the case of food contaminants and also in air pollutants. Examples are the association of carbon monoxide with nitrogen oxides, of carcinogenic polycyclic hydrocarbons with certain solvents (n-dodecane, for instance), or of tobacco smoke with asbestos (a worker in the asbestos industry who is also a smoker has 8 times more chance of dying of lung cancer than a smoker who smokes the same amount but has no contact with asbestos. The first man also has 92 times more chance of dying of lung cancer than his non-smoking fellow employees)—these are all well-known mixtures for increasing the toxic potential of the compounds in question.

Even possible synergism between chemical pollutants and ionizing radiation cannot be discounted *a priori*. Cook (1971) gives an account of an epidemiological study of a population of American uranium miners who were also cigarette smokers. The result was that the incidence of pulmonary carcinomas, which would normally affect an average of 15.5 cases per year in that particular socio-professional category (miners), was in fact as high as 60 cases in the mining population. It would seem from this study that there exists a clear potentiation between tobacco smoke and irradiation of the pulmonary parenchyma by the traces of radon and radium particles contained in the uranium mineral.

It appears, therefore, that the notion of a threshold limit value will always be marred by an unavoidable margin of error as long as it does not take account of the processes of synergism.

Interactions of Pollutants

There exists a fundamental miscalculation which leads to a systematic underestimation of the danger of the pollution to which we are exposed. It comes from the fact that in evaluating maximum tolerable doses we do not allow for the eventual summation of the effects of different micropollutants.

(a) Establishing maximum tolerable doses—Let us first re-examine how maximum tolerable doses are determined. This involves long-term experiments on laboratory animals. The animals are raised in a synthetic atmosphere containing a known concentration of an aeropollutant or are fed with food contaminated by given doses of the substance being tested. In almost all cases, the experiments concern only one toxic substance. By this method, an evaluation

is made of the weakest possible dose level of the toxicant that can cause long-term problems discernible with the aid of histological or other methods of testing. However, it is worth noting at this point that certain abnormalities observed by using an electron microscope, such as proliferation of the endoplasmic reticulum in hepatocytes due to the action of halogenated hydrocarbons, are not taken into consideration by homologation committees on the pretext that it is not possible to associate a precise pathological disorder with cytological changes of this nature!

Once the minimum dose is established, the maximum permissible concentration of the substance in the atmosphere or in foodstuffs is fixed by dividing the minimum dose usually by 50 or 100. In the case of food contaminants the whole operation is more complex, as it has to establish an acceptable daily permissible dose expressed in ppm which takes into account the type of foodstuff liable to contamination and the amount of the foodstuff ingested every day by an average individual. Fixing these dose levels, therefore, runs the risk of meeting serious methodological difficulties.

To begin with, these so-called long-term experiments rarely last more than two or three years, whereas a human being will be exposed to the various substances throughout his life. Secondly, certain population groups with a particular diet may in fact absorb much larger doses of the contaminants than those recommended by legislation. An example of this was observed in Cumbria where several thousand of the inhabitants were consuming large quantities of laver bread made with seaweed contaminated by the ^{106}Ru (ruthenium) discharged into an estuary by the fuel reprocessing plant at Windscale (after Foster *et al.* in Ramade, 1978/1982). The third and major criticism of this method of evaluating the benefit/risk relationship is that at no time is man ever exposed to one toxic substance only or to just a small number of them.

(b) The problem of the multiplicity of pollutants—In today's world food is being systematically contaminated by pesticide residues, by food additives (colourings, flavourings, stabilizers, emulsifiers, etc.), by traces of medical substances used in zootechny (antibiotics, sulphamides, etc.); the populations of industrialized countries are taking medicines which are prescribed too often without moderation (for instance, 1000 million (1 billion) tranquillizers were consumed in France in 1982); in addition, the air in cities is now a real cocktail of micropollutants (SO_2, Pb, nitrogen oxides, CO, asbestos dust, carbon black, etc.); and of course all populations are exposed to the X-rays from medical radiography and from our television screens, as well as to the radionuclides dispersed into the environment by commercial and military uses of the atom.

What ecotoxicologist could contend that the sum of all these hundreds of micro-doses, individually said to be 'tolerable', is still tolerable to man, who is constantly exposed to their total effects? If we just think of the enormous number of pollutants being inhaled and digested by the average city-dweller, we cannot fail to correlate this growing contamination with the increased frequency of degenerative diseases in the urban environment.

D—THE INFLUENCE OF ECOLOGICAL FACTORS ON THE EFFECTS OF TOXICITY

The action of toxic substances on living organisms is conditioned by the various ecological factors peculiar to each ecosystem and to man's environment. These factors can be divided into two groups: intrinsic factors of a biotic nature characteristic of the species being studied; and extrinsic factors, both biotic and abiotic, which define the particular conditions of the environment being studied.

I—The Influence of Intrinsic Factors

Taxonomic Variations

The unique biological characteristics of the species being studied are intrinsic factors, as are the particular features of a breed, race or family group, that is a group of individuals having a common pool of genes. Other intrinsic factors are the special characteristics of each ecophase (each stage of a life-cycle defined by differing specific ecological needs) and the physiological state (normal or abnormal) of each individual under study.

(a) Artificial pollutants—There is great variety in responses and sensitivity to pollutants depending on the taxonomic group and even between individual members of the same family group or other smaller systematic subdivisions of populations. In some cases, effects of the toxicity of a pollutant can be very selective on just one constituent element of a community. For example, the LD_{50} of phosphamidon, an organophosphate insecticide, varies from 1 to 30 according to the species being tested. Table 1.5 shows the comparisons between the toxicity of a few pesticides on four species of birds, of which three belong to the same order, the Galliformes.

In the same way, the toxicity of penicillin varies from 6 mg/kg for a guinea-pig to 1800 mg/kg for a mouse, in other words an LD_{50} which is 300 times greater for a mouse (Truhaut *et al.*, 1974).

Such differences in reactions and tolerance are also seen with cases of long-term toxicity by various toxic substances. TOCP (mentioned earlier, page 17), for instance, causes demyelination of the medullary and peripheral axons in

Table 1.5 The LD_{50} expressed in ppm of a few pesticides (absorbed by mouth) for different species of birds

LD_{50} Species	DDT (I)	Lindane (I)	Fenthion (I)	Phosphamidon (I)	Paraquat (H)
Colinus virginianus	611	882	30	24	981
Coturnix c. japonica	568	425	86	89	970
Phasianus colchicus	311	561	202	77	1468
Anas platyrhynchos	1869	>5000	231	712	4048

I = Insecticide   H = Herbicide

both man and the domestic hen, although it has no effect on several other species of primate (*Macacus rhesus* (rhesus monkey), for example) or on rodents. These variations in the response to toxic agents from one species to another have in the past actually caused serious errors in pharmacological tests. A classic example was the relative insensitivity of female rats to thalidomide, which then caused such terrible teratological disorders in the children of pregnant women who had used the drug as a sedative. If only the animal experiments had been carried out on female rabbits or mice the danger of the drug would in fact have been exposed immediately, as it has strong teratogenic effects on those species.

(b) Natural toxicants—There is also a great variety in the tolerance levels of different species to natural toxicants. α-Amanitine, one of the main poisons of the *Amanita phalloides*, is toxic to both man and mice at a dose level of 0.1 mg/kg, whereas rats have ten times more resistance to it (1 mg/kg). In the same way, the LD_{100} of phalloidin, the most common toxin in this fungus in terms of mass, is to the order of 3 ppm for a mouse but at least 100 ppm for Gasteropoda pulmonata. This explains the relative insensitivity of slugs to the *Amanita phalloides*, whose peridium often bears the traces of having been bitten by these molluscs (Heim, 1963).

Different levels of tolerance to natural toxic substances are also found in vertebrates. Among the various species of freshwater Teleostei, only the Salmonidae family, especially the trout, develops hepatoma in response to aflatoxin. Quails (*Coturnix coturnix* L.) are resistant to the alkaloid, cicutin, contained in hemlock, although birds which have fed on the substance become poisonous to the human consumer when the toxin reaches human muscles (Truhaut et al., 1974).

Plants, too, have different levels of tolerance to toxicants. For instance, certain phanerogams such as the plantain (*Plantago lanceolata*) or the stellaria or starwort (*Stellaria media*) can grow in an atmosphere containing more than 1 ppm of sulphur dioxide, although not a single lichen can survive prolonged exposure to 30 ppb of the same gas.

(c) The application of differences in tolerance levels—It is precisely on the varied responses to toxicity of different taxonomic groups that the use of herbicides is based. The diffusion of phenoxyacetic acid derivatives in cereal culture works on the principle of the Graminaceae having a stronger resistance to these substances, whereas weeds of the dicotyledon group are very sensitive to them.

The Role of the Ecophase

Tolerance to different pollutants also varies according to the stages in the life-cycle (ecophases) of the organism being studied. For vertebrates, the young and, of course, embryos are generally far more vulnerable than adults to the action of a particular pollutant. The same is true for invertebrates, whose larvae are usually more sensitive than the adults. This is probably due to there being greater metabolic activity in the young, growing organisms which causes them to absorb, mobilize and circulate to all body tissues larger quantities of the toxic substance for the same level of environmental contamination.

In contrast, maximum tolerance to pollutants is shown by dormant eggs protected by a thick chorion, or by the diapausant nymphs of holometabolous insects whose metabolism is greatly slowed down. For example, the chrysalids of a number of the Lepidoptera order have a strong resistance to hydrocyanic acid although this gas is extremely toxic to adult butterflies and moths. There are also significant variations in sensitivity during the imago stage of an insect's life. The adult house-fly has a resistance to insecticides that varies up to as much as fourfold with age.

The Role of the Family Group

Within a species of living organisms there are important variations in tolerance to polluting agents and other toxicants according to the genetic group—the family—being considered. This is particularly well known to entomologists or pharmacologists working with the lower orders of organisms such as bacteria or pathogenic fungi. There are many examples which reveal the enormous differences between tolerance levels that can separate various strains or familial groups within the same species—strains of bacteria with resistance to antibiotics and which can even become dependent on them, phytopathogenic fungi resistant to fungicides, insects resistant to insecticides, etc.

The study of these resistance mechanisms constitutes an important field for ecotoxicology and can provide essential information for the understanding of ecological genetics, of processes of adaptation and eventually of the fundamental mechanics of evolution.

Some strains can of course have the opposite reaction and be hypersensitive to certain toxic agents. The notion of idiosyncrasy or congenital intolerance explains such a reaction, which is of an hereditary nature and due to the deficiency of a genome, as opposed to an allergic reaction which is an acquired intolerance and not genotypic. In most cases, the idiosyncrasy results from a lack of one or more of the enzyme systems capable of metabolizing the substance causing the hypersensitive reaction. Well-known cases of intolerance in human toxicology are those of certain ethnic groups in the Mediterranean area to natural substances contained in the common broad bean, and the sensitivity of white races to nitrated phenol derivatives whereas both black and yellow races are more resistant. American blacks have shown particular sensitivity to an anti-malarial drug, primaquine, because of deficiency of an enzyme, glucose-6-phosphate-dehydrogenase, which intervenes in the pentose pathway. Similarly, lactose intolerance in adolescents and adults of various ethnic groups in tropical African regions results from the inability of their intestine to synthesize lactase, the enzyme responsible for hydrolysing this diholoside.

Adaptation to Toxic Substances

Permanent exposure of a species of living organisms to a toxic substance leads, as a general rule, to a process of adaptation. This means that there will be a spectacular rise in the response threshold to the toxicant at the individual level and also at the level of the population and even of the whole species. In this way a true process of selection takes place of pollutant-tolerant ecotypes resulting sometimes in the appearance of strains with such a strong resistance to a toxicant that they can apparently withstand easily concentrations of the substance that are several times greater than those causing 100% mortality in sensitive populations.

The term resistance coefficient, Fr, to a toxicant is given to the relation between the LD_{50} of the strain resisting the toxicant and the LD_{50} of a sensitive reference strain: $Fr = LD_{50}(R) / LD_{50}(S)$. Resistance coefficients can reach very high values. Some of our own research has shown that for a strain of house-fly (*Musca domestica*) which had become resistant to insecticides, the coefficient was more than 10 000 for dieldrin, an insecticide of the chlorinated cyclodiene group (Fig. 1.12).

It is now very common for natural populations to be resistant to most of the groups of toxic chemicals used systematically by man in both urban and rural environments. The resistance of bacteria to antibiotics, of fungi to fungicides and of some weeds to herbicides is already posing difficult problems

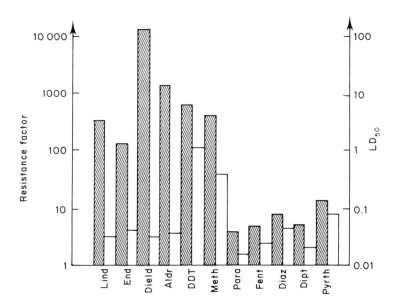

Fig. 1.12 Chart of the resistance of a strain of house-fly (*Musca domestica*) to the main types of synthetic insecticide stripped bars = LD_{50} of resistant strain, white bars = LD_{50} of sensitive ones. (Lind = Lindane, En = Endrin, Dield = Dieldrin, Aldr = Aldrin, Meth = Methoxychlor, Para = Parathion, Fent = Fenthion, Diaz = Diazinon, Dipt = Dipterex, Pyrth = Pyrethrin) (After Ramade, 1967.)

Fig. 1.13 The conversion of DDT to DDE by the DDT-dehydrochlorinase

by undermining the efficiency of these biocidal agents. In the phylum Arthropoda, the growth in the number of species that have adapted themselves to insecticides and other pesticides is particularly notable. In 1965, 182 species of insects and mites were counted as being resistant to pesticides; by 1970 this had risen to 250 species, and to more than 400 species by 1980.

It has been virtually established that with any species of organisms exposed to a chronic level of sublethal concentrations of a pollutant, a process of adaptation will occur. By adapting, the individuals of a population living in a contaminated environment are able to survive, without any apparent ill-effects, doses of the substance which would cause physiological damage to any individuals who had never yet been exposed to that particular toxicant. The speed and scale of the adaptation do, however, vary considerably with the nature of the toxicant and the species involved. The process is more rapid in more primitive organisms with a greater biotic potential. Thus, a strain of bacteria will become more readily resistant to a toxic agent than will a species of insect. Nevertheless, in spite of it being a longer and more difficult process, experiments have shown that it is even possible for a resistant strain of mammals to appear (e.g., the resistance of rats and other domestic rodents to anticoagulants).

The adaptation of any plant or animal species to a toxic agent is conditioned by the existence of deactivation mechanisms, and in particular by the presence of adaptable enzymes in the organs of the target specimen capable of breaking down and neutralizing the contaminant. In the case of DDT, for example, it was shown that resistance to the insecticide generally resulted from the synthesis in different tissues of the insects (especially fatty tissues) of a detoxicating enzyme, DDT-dehydrochlorinase which transforms DDT to DDE (Fig. 1.13).

This DDT-detoxicating enzyme is also present in sensitive as well as resistant strains, but it occurs in far greater concentrations in the tissues of the resistant strains. Furthermore, it has been established that a strong correlation exists

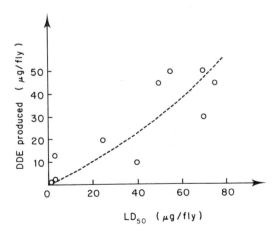

Fig. 1.14 Correlation between the level of resistance to DDT of various strains of house-fly and the level of their production of DDE. (*in* O'Brien, 1967.)

between the level of DDT-dehydrochlorinase activity in an insect strain and the degree of resistance to DDT shown by that strain (Fig. 1.14).

There are numerous other examples of the adaptation of animal and plant species to the pollutants present in the environment. In an aquatic ecosystem, for instance, it was shown that phytoplanktonic algae (*Skeletonema costatum* and *Amphidinium carteri*) can develop a resistance to the toxic effluents from pulpmills into coastal waters (Stockner and Antia, 1976). Whereas these diatoms only tolerated weak concentrations of the diluted effluents initially, exposure of the same strain of phytoplankton to gradually increasing concentrations of the pollutants resulting in a significantly higher tolerance threshold beyond which the toxic effluent inhibited the growth of cultures of these species.

Evidence has also been found of adaptation mechanisms to the toxicity of heavy metals contaminating soil or water. An example is the research conducted by Bryan and Hummerstone (1971) on populations of polychaeta Annelida (*Nereis diversicolor*) in various estuaries in South-West England where the sediments were polluted by effluents from the copper mines and by other toxic metals. Comparison between tolerance levels to copper of the strains of *Nereis* sampled from contaminated biotopes with those from pollution-free estuaries showed clear adaptation to the presence of these heavy metals. The LC_{50} of copper reached 2.5 μg/cm^3 of sea-water after 24 hours in the Annelida from contaminated estuaries containing 4000 μg of copper per gram of sediment, whereas the LC_{50} was below 0.5 μg/cm^3 for strains of *Nereis* taken from non-polluted estuaries (that is, with less than 20 μg of copper per gram of sediment).

The resistance of living organisms to heavy metals is generally associated with processes of deactivation. Adaptation of marine organisms to these substances seems to depend particularly on the synthesis of metallothioneins (cf. for example, Noel-Lambot, *et al.*, 1980). Common eels (*Anguilla anguilla*), for instance, when exposed to sublethal concentrations of mercuric chloride, become tolerant to doses of the compound which would normally be fatal. Their adaptation is accompanied by a massive rise in the level of metallothioneins tolerated by the animal's different organs.

The Effects of Adaptation on Dose–Response Curves

The appearance of strains of organisms with a resistance to toxic substances can have various effects on dose–response curves.

It is accepted that the probit–dose curves occasionally have their slope modified by successive generations of organisms. In a first phase, the slope is reduced progressively as the LD_{50} of the strain increases in relation to the gradual elimination of the more susceptible phenotypes. Subsequently, the increased homogeneity of the population as regards their resistance level will be marked by a significant rise in the slope of the curve.

In other cases, the increase in the LD_{50} occurs in successive generations without modifying the slope of the ld – p curves. This is because the adaptation will result from a general rise in the strength of all the individuals in the exposed population, rather than from the selection of pre-adapted resistant phenotypes (Brown, 1978).

Variations with Morbidity

There is a whole group of factors affecting pathological conditions and the activity of toxicants. Organic differences caused by various diseases can increase sensitivity to polluting agents. Examples are hepatic illnesses which necessarily affect the individual's detoxication capacity, or kidney damage which will mean less resistance to toxic substances. It also seems that thyroid and adrenal cortex deficiencies lower tolerance levels to the effects of pollutants. Malnutrition will also reduce tolerance levels.

Conversely, certain toxic compounds can actually increase the receptiveness of organisms to infection by various pathogenic agents. For instance, the presence of insecticides in the soil stimulates the strength of infection by myocoses caused by *Beauveria* and *Metarhizium* for scarabaeid beetle larvae.

An experiment demonstrating this interaction between toxicants and agents of infection has been done on human hepatic cells. When DDT is added to *in vitro* cultures of hepatocytes it increases their sensitivity to various viruses.

II—The Role of Extrinsic Factors

Whatever their toxicity, the different pollutants released into the natural environment will be exposed to the action of abiotic ecological factors (temperature, water, light, etc.) and biotic factors (degradation by micro-organisms). Concerted microbial action will tend to neutralize the pollutants by transforming them into less toxic derivatives.

The Natural Activation of Pollutants

Under certain circumstances, pollutants can in fact react with biogeochemical factors to form compounds which are equally as toxic or even more toxic than the original substances. Parathion, for example, a powerful organophosphate insecticide, can be oxidated through the influence of abiotic processes or biological systems, producing the following general pattern:

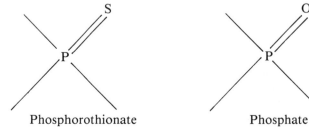

Therefore, when an animal absorbs parathion, the substance is transformed by certain cellular oxidases to paraoxon which is far more toxic. In fact the LD_{50} for oral administration of parathion is 7 mg/kg for a rat but the LD_{50} of paraoxon is only 0.7 mg/kg. When parathion enters natural waters it is slowly hydrolysed and in turn the products of hydrolysis, by a secondary reaction, form pp'dinitrophenol, a fairly stable substance which is very poisonous to various aquatic animals.

The Activation of Pollutants by Interaction

Pollutants can of course combine and react with each other to form substances that are more toxic than the original single toxicant. For instance, peroxyacyl-nitrates (PAN) form in atmospheres heavily polluted by nitrogen oxides and hydrocarbons if there is a sunny climate. In the first phase of this, the ozone in the atmosphere reacts with the hydrocarbons and forms peroxyacyls:

$$R-\underset{\underset{O}{\|}}{C}-O-O- \tag{17}$$

Then, in a second phase, the following reaction will occur:

$$R-\underset{\underset{O}{\|}}{C}-O-O- + NO_2 \rightarrow \underset{\underset{\underset{PAN}{O}}{\|}}{R-C}-O-O-NO_2 \tag{18}$$

The PAN compounds are, therefore, contaminants produced by the interaction of primary pollutants, and with all other things being equal, their toxicity can actually be hundreds of times greater than the original pollutants from which they were derived.

Interaction between Bacteria and Pollutants

Micro-organisms represent an extremely important biogeochemical factor. They can degrade most chemical pollutants, even *a priori* stable compounds, and transform them generally into less toxic substances, although the products of degradation can be as dangerous, if not more so, than the initial pollutant. It has been shown that DDT could be transformed by edaphic bacteria into acetonitriles, which are fairly poisonous compounds. Also, Jensen and Jernolov (1969) have demonstrated the fundamental role played by benthic bacteria in the conversion of mineral mercury and organic mercury compounds to methyl mercury, an extremely toxic substance.

Interaction with Abiotic Factors

The abiotic factors in the environment (temperature, hygrometry; pH and dissolved oxygen capacity in aquatic ecosystems) are of primary importance in the effects of pollution. The toxicity of pollutants to cold-blooded animals, to all aquatic organisms and even to warm-blooded animals to some extent is conditioned by abiotic factors. Temperature, for example, plays a major role in the action of insecticides on insects. Depending on the substance, the temperature coefficient can be either positive or negative.

The Influence of the Chemical State on Toxicity

The chemical state of a substance largely determines its toxicity. This factor conditions in particular the toxicity of heavy metals occurring either naturally or artificially in the environment. The saturation level of the valencies, the proportion of the compound which is in a dissociated state—linked with the pH of the environment and the nature of the soil being studied, and also the fact that an element can appear in the form of a mineral or as a compound with organic substances—all are factors which can greatly alter the toxicity of a given simple substance at a 'nominal' concentration.

For heavy metals, the example of chromium illustrates perfectly the role of the saturation level of valency in the degree of toxicity. The threshold level of toxic concentrations of hexavalent chromium for animals is in the order of 0.1 mg/litre of water, whereas the threshold for trivalent chromium is approximately one gram per litre. As a general rule, the complexing of heavy metals by organic compounds diminishes their toxicity.

In natural waters, the amount of dissolved organic matter can affect the toxicity of a pollutant by reducing its effective concentration. Various studies have shown that the complexing ability of continental waters is quite high, for instance in Lake Ontario, this ability is 1.23 $\mu M/l$ for divalent copper (from Förstner and Wittmann, 1981).

Studies using synthetic complexing agents have shown that these substances are capable of combining by chelation to trace amounts of metals present in natural waters. This is the case with NTA (Shaw and Brown, 1974) and with EDTA (Gardiner, 1976).

In general, this complexing of heavy metals by natural or synthetic substances reduces their toxicity. In the rainbow trout (*Salmo gairdnerii*), NTA significantly diminishes the toxicity of both zinc and copper, with the average survival time of the trout being increased by a factor of 10 for a concentration of 5 ppm of copper in water (cf. Fig. 1.15).

However, chelating agents can also produce negative effects in the natural environment, because these complexes can themselves present an intrinsic toxicity above a certain level of concentration. In addition, they can remobilize toxic mineral substances accumulated in sediments. In fact, an important fraction of the heavy metals present is combined with the colloid matter in the benthic

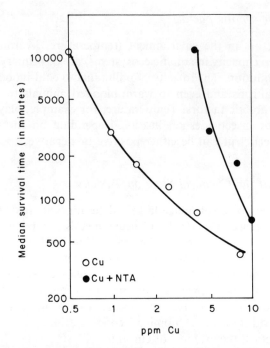

Fig. 1.15 The influence of a chelating agent, NTA, on the toxicity of copper for the rainbow trout (*Salmo gairdnerii*). (After Shaw and Brown *in* Förstner and Wittmann, 1981.)

silts (clay–humus colloidal complex). Batley and Gardner (1978) showed that in an Australian estuarine zone, 40–60% of the copper, 45–75% of the lead and 15–35% of the cadmium are combined in this way in the sediments.

The pH of soil and natural waters, which is largely conditioned by the amount of organic matter they contain, can also influence the degree of adsorption or desorption of the cations. In this way, the presence of amino acids both in open waters and in the sediments of limnic or coastal distrophic environments is capable of mobilizing toxic cations by desorption or on the other hand of neutralizing them owing to their amphoteric properties.

E — ANALYTICAL METHODS FOR DETECTING POLLUTANTS

Studying the contamination of various habitats and of human beings by pollutants raises considerable analytical problems. The detection of toxic substances present in concentrations of less than 1 ppm, and frequently of the order of 1 ppb (part per billion) requires very elaborate microanalytical techniques which go far beyond the scope of this book.

The discovery of gas chromatography with the capture of electrons and its application to the study of organohalogen compounds enables traces of these

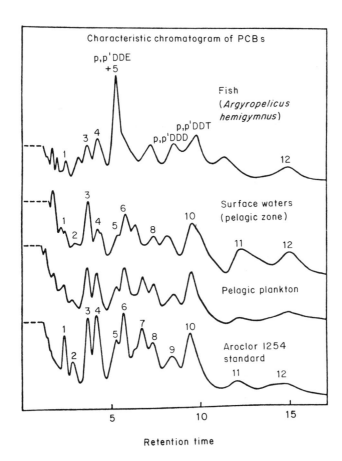

Fig. 1.16 Gas chromatograms of DDT and polychlorinated biphenyls (PCBs). In addition to the peaks of DDT and its derivatives (DDE and DDD) and be seen 12 peaks corresponding to the PCBs. These diagrams were obtained by analysing samples of seawater and of various organisms (plankton, fish) obtained in the pelagic zones of the central Atlantic. At the bottom is shown the chromatogram of a PCB (Arochlor 1254) which serves as a reference. (After Harvey *et al.*, 1974.)

substances to be detected, even in such minute concentrations as 1 ppt (part per trillion), or 10^{-12}. This technique has been used, along with other successful methods for identifying major pollutants, to distinguish polychlorinated biphenyls from organochlorine insecticides, whereas these substances would be virtually indistinguishable one from the other using any other analytical technique.

In another field, flame emission spectroscopy and atomic absorption spectroscopy are extremely useful in the detection of heavy metals and other mineral pollutants. Another analytical technique of interest to ecotoxicology is thin-layer chromatography, which is being used more and more for detecting natural toxic substances and various micropollutants.

These analytical methods cannot be elaborated here and the reader is advised to consult the considerable amount of specialist literature on the subject.

F—THE MONITORING OF POLLUTANTS

During the last ten years, various national and international authorities have undertaken to monitor pollutants in the environment, particularly those contaminating the atmosphere, continental waters and the oceans. An efficient system for managing and conserving natural resources or for protecting man's environment, faced with the growing number of pollutants and the importance of waste products, needs to gather a considerable amount of information on:

(a) the nature of potentially dangerous chemical substances;
(b) the sources, quantities and dispersal of pollutants;
(c) the ecotoxicological effects of these substances;
(d) the eventual trends in evolution of the concentrations and effects, as well as the causes of these changes;
(e) the means by which these waste products, concentrations, effects and trends can be modified and at what cost.

The role of monitoring in the series of events which have occurred since the problem was recognized until a decision was taken to control it has been analysed by SCOPE (1977) in a report inspired mainly by the work carried out at the Monitoring and Assessment Research Centre (MARC) in Chelsea (Holdgate, 1979).

At the end of the 1970s, under the aegis of the PNUE, the International Register of Potentially Toxic Chemicals (IRPTC) was established. The objective of this organization is to set up a 'bank' of all available data from all over the world on the various toxic compounds known to be environmental pollutants or prone to become pollutants if their usage becomes commonplace. The aim of the IRPTC is to catalogue the chemical properties of potential toxic substances, their analytical methods, their biological activity, their ecotoxicological effects and also their principal uses and the legislative provisions governing their use.

Earlier, at the end of the 1960s, the urgent need had been acknowledged for systematic monitoring programmes for major pollutants in the principal macroecosystems, bearing in mind the scarcity of basic analytical data available at the time for the main groups of pollutants, even in developed countries. This need was underlined for the first time as being of global urgency by the group for 'Study of Critical Environmental Problems' organized under the aegis of the MIT (cf. for example, Matthews *et al.*, 1971), and gave rise to several monitoring programmes in the United States. The Nationwide Pesticide monitoring programme established during 1965 in the USA was probably the first venture whose objective was to study systematically all pollution right across a whole sub-continent. Within its framework, the Bureau of Sport, Fisheries and Wildlife initiated an enormous monitoring operation of the organochlorine residues in waterfowl, taking as its reference sample the ends of the wing-tips of mallard ducks and black ducks (*Anas platyrrhynchos* and *A. rubripes*) sent to the researchers by hunters from all the different American states (Heath, 1969). The programme was later extended to the study of lead (Stendell *et al.*, 1979).

Perhaps the most systematic monitoring programme ever started is the 'Mussel Watch' (Goldberg, 1975 and Goldberg, *et al.*, 1978). Initially, it concerned the whole length of the Atlantic and Pacific coasts of the USA and was set up with the support of the EPA. The choice of lamellibranchs — mussels (*Mytilus* sp.) and oysters (*Ostrea* sp. or *Crassostraea* sp.) — as the reference species for the detection of major micropollutants in the

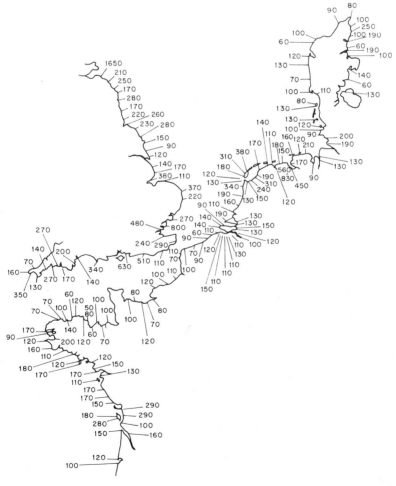

Fig. 1.17 An example of the application of the 'Mussel Watch' to the monitoring of contamination by mercury of the coastal waters of western Europe. The figures refer to ppb of soft-body wet weight. (After Wolff *in* Gerlach, 1981.)

ocean environment is due to their microphagous diet. These molluscs filter considerable quantities of sea-water and, therefore, accumulate pollutants more easily, playing in this way the role of 'sentinel' organisms. The object of the 'Mussel Watch' is to evaluate the chemical and nuclear pollution of marine coastal waters, with analytical detection of the amounts found in the mussels and oysters of heavy metals, organochlorine compounds, hydrocarbons and radionuclides. The 'Mussel Watch' principle has also been applied to the coasts of western Europe. Fig. 1.17 presents as an example the results obtained for mercury.

Whatever the interest in discovering evidence of residues of polluting substances in living creatures belonging to a particular ecosystem, this can be only a preliminary step in any programme monitoring environmental pollution. The most important aspect of this work is in evaluating the physico-chemical and especially the ecotoxicological impact of concentrations of pollutants shown up by the systematic analyses.

During the last few years the attention of specialists has been focused on the problem of **bioassays in the monitoring of pollution** of terrestrial and aquatic ecosystems. This whole question was analysed particularly thoroughly by the International Council for the Exploration of the Sea, whose symposium proceedings were edited by MacIntyre and Pearce (1980).

Certain specifications are laid down for the necessary techniques used to establish bioassays (Stebbing et al., 1980). These imply the following:

— The choice of test organisms of ecological and/or economic importance,
— The availability of the species being tested throughout their annual cycle,
— The existence of a biotest technique that is easy to use, not very costly and requiring little equipment,
— The homogeneity both morphologically and physiologically of the organisms either bred or gathered in their natural environment for the experiments, so as to minimize variations in susceptibility to the toxic substance. For this, it is strongly recommended to choose clones or homozygous strains where available,
— The experimental data obtained should satisfy usual statistical criteria.

The aim of using bioassays in the monitoring of environmental pollution is to establish a relationship between the toxicity and the concentration of a pollutant being studied in the biotopes concerned. A rational sequence of toxicity tests starts by determining the lethal effects on a short-term basis and continues with more sophisticated tests of longer duration designed to show the more subtle physiotoxicological and ecological effects. This sequential testing procedure provides a gradually more precise estimate of the minimum concentration of the pollutant likely to induce harmful ecotoxicological effects (Fig. 1.18).

Alternatively, it can be used to establish the maximum acceptable concentration of a pollutant in a given environment, without the deliberate

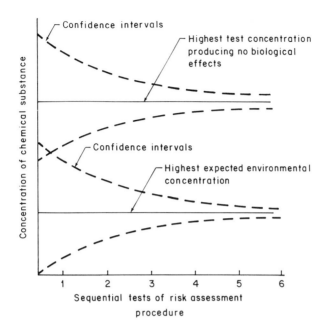

Fig. 1.18 A diagrammatic representation of the sequential procedure for assessing the risk involved with a chemical substance showing clearly the gradual narrowing of the confidence interval in evaluating the minimum acceptable concentration. (After Cairns, 1980.)

application of the chemical causing any unfavourable biological consequences (Cairns, 1980).

The sequence of tests chosen is conditioned by the effects and responses corresponding to the biological monitoring being considered. Wilson (1980) proposes a series of criteria to which the selected effects must conform for the use of biotests (Table 1.6).

A large number of animal and plant species have been selected as good material for biotests (Stebbing *et al.*, 1980). In aquatic environments, along with various species of phytoplankton and the larvae or adults of bivalve molluscs, numerous species of invertebrates can also be used as reference species in the monitoring of the biological quality of water: echinoderm larvae, cladoceran crustacea, copepods and Mysidacea, hydra and the fry of fish. In general, the essential criteria for choosing test species should be to have stenoecious species. It is a universal ecological fact that stenoecious species are more vulnerable to any modification of their environment than are euryoecious species.

To end this brief look at the problems relating to bioassays in the monitoring of environmental pollution, it is worth noting the methodological gaps which still exist in the field of evaluating synecological effects. Indeed, if it can be considered that the methods for testing the effects of a given pollutant on an individual or population might be relatively reliable, even when done over a

Table 1.6   Assessing the suitability of effects and responses for biological monitoring. (After Wilson, 1980.)

The questions below are pertinent to the selection of effects; others could no doubt be added. Initially it is important to formulate the objective of the monitoring exercise precisely in order to determine if biological monitoring is an appropriate technique. It is necessary to formulate the target, the geographical scope, location and time base. Once these have been established the following questions could be structured as a decision tree. This has not been attempted.

1. Can the objective be monitored directly?
2. Is the objective (or substituted effect) related to exposure?
   Is this relationship quantified?
3. Is there a time lag between exposure and effect on the objective?
   Is the time lag significant for maintaining the objective?
4. In selecting an effect that occurs at lower exposure, is the objective functionally related to the effect, and what is its significance?
5. Is it possible to relate the selected effect to other effects, at the same or different levels of organization?
6. Is the effect reversible, and over what time scale?
7. How specific is it with respect to stressors?
8. How sensitive is it, i.e., what degree of exposure is required to elicit it?
9. How responsive is it, i.e., how great a change in the degree of exposure is required to cause an observable response!
10. Is it variable, and what is its normal or acceptable variability?
11. Is it clearly discernible?
12. Can it be quantified, and with what accuracy and precision?
13. How easily can it be quantified, i.e., is it a practical proposition?
14. What is the extent of the practical and theoretical knowledge to support its use?

long period of time, it is certainly not the same thing when it comes to studying a whole community. Much progress is also needed in improving the tests relating to phenomena of bioconcentration and biomagnification. Obviously, the definition of quantitive criteria which will enable the evaluation of the impact of a toxic substance on the structure of a population and of the biocoenosis still remains today a top priority for ecotoxicology.

CHAPTER TWO

The Pollution of the Biosphere

A — Pollution	59	C — The classification of pollutants	70
I — Definition	59	D — Dispersion and circulation mechanisms of pollutants	71
II — The history of pollution	60	1. Atmospheric transport of pollutants	71
B — The causes and significance of the pollution of the biosphere	61	2. The transfer of pollutants from the atmosphere into water and soil	75
I — General points	61	3. The transfer and concentration of pollutants in the biomass	77
II — The principal sources of pollution	62	4. The transfer and concentration of pollutants in the trophic chains	80
1. Energy production, an essential source of pollution	62	5. Conclusion	85
2. The modern chemical industry, source of various pollutants	66		
3. Modern agriculture	68		

Ecotoxicology can be defined as the science whose object is to study the ways that pollutants are dispersed into the biosphere and the mechanisms by which different ecosystems are contaminated. It also aims to show the interaction of pollutants with biotic and abiotic ecological factors and to determine their effects on living beings at all levels — that of the individual, the population and a whole community.

A — POLLUTION

I — Definition

The term 'pollution', which is used extensively these days, covers all the waste toxic chemicals released by man into the ecosphere, as well as substances that may not be particularly dangerous to organisms but which have a disrupting influence on the environment.

Etymologically, to pollute means to profane, defile, dirty, degrade; such synonyms are unequivocal and are just as meaningful as the long definitions given by experts. Of these, there is one worth mentioning, published in a report prepared in 1965 by the official scientific committee of the White House, entitled 'To restore the quality of our environment'. The report declares:

> Pollution is an unfavourable modification of the natural world which appears either entirely or in part to be a by-product of human action, changing through direct or indirect effects the criteria governing the distribution of energy fluxes, radiation levels, the physico-chemical composition of the natural world and the abundance of living things. These modifications can affect man directly or through agriculture, water and other biological products. They can also affect man by altering the physical objects he possesses, the recreation possibilities of his environment, or even by destroying the beauty of nature.

This definition includes in fact all of man's actions which destroy the biosphere. But it is also important to include, apart from artificial pollutants, those toxic substances which occur naturally but whose volume is increased by man. In this category can be grouped the aflatoxins, various bacterial toxins used in the food industry, the microbiological pollution of water, etc.

II — The History of Pollution

The history of pollution faithfully reflects the progress of technology. The first causes of contamination of the environment occurred in the Neolithic period. In that period, the discovery of agriculture meant that human groups became settled, leading to the creation of towns where for the first time the population density exceeded by a long way that characterizing any other species of mammal, even the most gregarious. However, these sources of pollution remained very limited. They came from the microbiological contamination of water by domestic effluents and more rarely from the treatment of various toxic non-ferrous metals by primitive methods.

Throughout the intervening centuries and right up until the beginning of the industrial era, in eighteenth-century Europe, pollution was very limited. It was the birth of large-scale industry, in the middle of the nineteenth century, which saw the contamination of water, air and sometimes even soil becoming a local concern in the areas round mining or metallurgical installations and in the large overpopulated industrial cities.

Whatever the importance of pollution problems up to the Second World War, none of them ever posed such a disturbing threat as modern technology and the growth in recent decades of industrial and urban effluence, the accumulation of waste materials, the release of new and incredibly toxic substances into the air, water and soil.

The most serious questions with which we are now faced in ecotoxicology arise from the release into the environment of substances which are both very toxic and also virtually non-biodegradable, if not indestructible.

Adding to the ancient causes of pollution, whose extent has increased in an alarming way, there are now new sources resulting particularly from the development of organic chemistry and the nuclear industry.

B—THE CAUSES AND SIGNIFICANCE OF THE POLLUTION OF THE BIOSPHERE

I—General Points

The vast majority of problems associated with our industrial civilization come from the fact that the natural energy flow has been disrupted by this civilization which has also broken the cycle of matter by producing growing quantities of non-biodegradable and, therefore, non-recyclable waste.

At the present time, both population and pollution are increasing at an accelerating rate, whereas the self-cleaning ability of the ecosphere is being reduced more and more by the dispersion of toxic residues and varies inversely, tending to become completely neutralized. In addition, the wastefulness of western countries whose present sociological structures urge frequent renewal of consumer goods, not only when they are worn out but also when they are out of fashion, is also combining to increase the scale of pollution by vast amounts. Nowadays, wear and tear is incorporated from the assembly line. It is even impossible to replace worn parts because commercial models are soon discontinued, forcing the consumer to renew completely the goods, even if he is unwilling to accept mass-media advertising.

In this way the volume of surplus products thrown on the dump is artificially inflated by obsolete consumer goods, a system created by our society. This scandalous squandering of energy and raw materials will lead humanity sooner or later to an irreversible shortage of basic materials for industrial and agricultural activity if the process is not stopped very soon indeed.

However, there are factors of a sociological nature which aggravate environmental pollution in the industrialized nations. Rapid urbanization is probably one of the most important. At the beginning of this century more than half of the population of Europe lived in the countryside—except in Great Britain. This was the case in France until 1939, but today 80% of France's population live in cities. The rural exodus became greater over the last decades, depopulating the provinces in favour of large metropolitan areas. The Paris region alone, which is being extended by the policy of building new towns, now numbers 10 million inhabitants which is twice the population of the whole of Finland! The largest metropolitan area in the world extends down the east coast of the United States from Boston to Washington and has a current population of nearly 40 million!

The concentration of industries and habitation in the same fairly restricted areas as regards levels of population density simply multiplies the problems for so-called developed countries. Pollution phenomena are definitely intensified mainly by accelerated urbanization and anarchic industrialization. This trend is amplified by the incessant reduction of woodland zones on the periphery of urban areas, zones which are fought over as prime sites for frantic housing development while their indispensable role in purifying the atmosphere should make them sacrosanct.

The combination of these different factors has resulted over the last decades in pollution growing more rapidly than population in the industrialized nations. In the United States, for example, the pollution index *per capita* of the population increased by 1000% between 1946 and 1970 while the population went up by 'only' 46% over the same period.

II—The Principal Sources of Pollution

The growth of pollution of the ecosphere is both quantitive and qualitative in character. In addition to the increase in production and consumption *per capita* created by technological progress, there is also the incessant multiplication in nature of the contaminating substances released by man into the natural environment. For example, modern organic chemistry permits the annual commercialization of several hundred new substances, often very toxic to living beings, but which are produced on a large scale of many thousands of tonnes per year before any study is made of their toxicological or ecological properties.

There are three main causes of pollution of the ecosphere by industrial civilization: energy production; the activities of the chemical industry; and agricultural activity.

For each of these fundamental causes of pollution there are basic sources at the production level at one end, and at the consumer level at the other.

1. Energy production, an essential source of pollution

In these days of energy crisis the western consumer, whether he be individual or legal entity, is beset by the problem of energy resources and their cost. However, for the ecologist the crisis started long before the general public or politicians ever became obsessed by the possible shortage of oil products. One of the essential elements of the energy crisis, taken in its widest sense, lies in the multiple pollution caused by its various uses in contemporary civilization. The insatiable hunger for energy which seized the industrialized nations, apart from leading to a frenzied waste of scarce and irreplaceable energy resources such as petroleum oil, plays a very important role in environmental contamination by countless toxic substances.

Fossil Fuels

(a) Coal—From the eighteenth century onwards, when coal began to be used

as fuel by city-dwellers and industries, the first atmospheric pollution was seen. The famous London smog ('pea-souper') was one of the best-known examples of this. During the nineteenth century, especially after 1860, the extraction of coal and petroleum took place on an ever-increasing scale to cater for the energy needs of big industry and to fuel the new methods of transport: railways and steam ships. In 1900, coal met 90% of world energy requirements as opposed to only 4% met by petroleum.

Since then, natural gas has been added to the list of fuels, while the use of petroleum has continued to increase to the detriment of coal. Thus, from 1929 to 1971, world coal production only rose by 70% whereas that of petroleum went up by 1000% in the same period. In France, for instance, coal accounted for 51% of national energy needs in 1950 but only for 18% in 1981. At the present time, hydrocarbons provide for 73% of energy consumption in western Europe. In the US, coal accounted for 80% of national energy consumption in 1925 and for only 19% in the late 1970s.

(b) Petroleum—To these qualitative changes in energy supplies must be added the incredible growth in energy consumption. In the United States, this has increased 70 times in just one century. In France, petroleum imports have grown from 5 million tonnes in 1939 to 109 million tonnes in 1978. Oil consumption increased in the USA from 9×10^6 barrels per day in 1960 to about 14×10^6 barrels per day in 1980. On a world scale, petroleum production has gone up by 285% from 1960 to 1979, when production reached 3.2 billion tonnes. However, since the end of the 1970s there has been a clear trend to reduce petroleum consumption in western Europe, due to the development of nuclear power, energy conservation policies and also because of the economic recession.

Another statistic which demonstrates the importance of the colossal energy consumption of our civilization is that the total sum of fossil fuels burned in 1969 was equal to 5% of the gross annual primary production from photosynthesis in the whole of the biosphere.

We are living today in the petroleum era, although its use as a fuel seems like a sacrilege comparable to the one committed by our ancestors who destroyed our forests in order to make charcoal. The use of fossil hydrocarbons enters every sphere of activity in our society, from industrial production (factories, thermal power-stations), down to cars and domestic usage. In fact, petroleum and natural gas are used primarily as fuels in spite of the unparalleled squandering of non-renewable resources that this represents. In France, 35% of the petroleum imported in 1972 was burnt as industrial fuel, 25% was used to heat commercial and domestic buildings, 22% was consumed as motor-fuel (petrol, etc.) and only 7% was used as a raw material in the chemical industry for various organic syntheses. During the early 1980s, 29% of the US fossil fuels demand was consumed in transportation, 38% was burnt as industrial fuels and 33% for household and commercial usage.

These facts may seem irrelevant in the context of this book. However, they do serve to underline the fundamental importance of the production and use

Table 2.1 Principal causes of pollution associated with the use of fossil hydrocarbons

Activity	Cause of pollution	Polluted environment	Nature of pollutants
Extraction	Leakage from offshore well(s)	Ocean	Crude oil
Transport	Accidents, 'gas extraction'	Ocean	Crude oil
Refining	Release of gaseous and liquid effluents	Atmosphere Continental waters Seas	Various organic SO_2 compounds, mercaptans, etc.
Utilization	Incomplete combustion	Atmosphere	SO_2 oxides of nitrogen hydrocarbons

of energy from fossil fuels in the pollution of the ecosphere. At all levels of human activity, their use places them in the first row of sources of environmental pollution (cf. Table 2.1).

(c) Consequences of the growth in energy demand—The extraction and use of petroleum is accompanied by numerous polluting effects as well as other ecologically absurd actions. Oil slicks and other oil leaks are contaminating the world's oceans, while oil refining is polluting continental waters. The building of refineries ruins many coastal sites and rich agricultural land, and the choice of their location may sometimes lead us to question the common sense of certain 'planners'.

This is particularly the case with the insane development of road freight transport and urban traffic control, all of which are devourers of space and fuel. Slowly but surely, all sorts of express roads, fly-overs, and motorways are devastating the urban environment and encroaching continuously on the rural zones, shrunken forest areas and rare small islands of nature which have been miraculously conserved from the old Europe. In addition, the proliferation of motor cars enables city dwellers, who were previously kept, half-deliberately, in total ignorance about the problems of nature conservancy, to transform gradually the few scarce areas of biological interest still surviving into public dumping grounds.

Elsewhere, the thirst for energy in industrialized nations is accompanied by an incessantly growing contamination of the atmosphere, continental waters and oceans by countless pollutants produced from the burning of fossil hydrocarbons. This pollution of the atmosphere and soil is gradually destroying the forests near urban areas. It is also menacing human health at a time when systematic research into making maximum profit seems to take precedence over any consideration for public welfare. An example to support this is the series of delays caused by the oil companies to removing the lead from petrol or to significantly reducing the sulphur content of fuels. You only have to consider

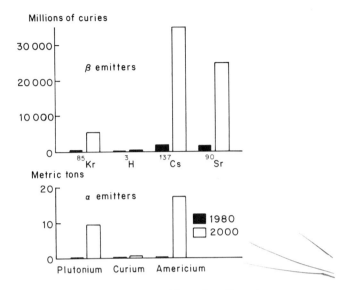

Fig. 2.1 The increase in quantities of radioactive waste produced by the nuclear industry in western Europe from now until the end of the century (OECD document). The β emitters with an intermediate half-life (^{85}Kr 10.5 years, tritium 12.4 years, ^{137}Cs 32 years, ^{90}Sr 28 years) are separated from the β emitters with a long half-life (nearly 25 000 years in the case of plutonium)

the dangerous epidemiological consequences of sulphur pollution in urban areas, especially the continual rise in chronic bronchitis.

Nuclear Power

New concerns have appeared over the last few years, magnified by the energy crisis in western countries. These concerns arise from the accelerated development of the nuclear industry. Apart from the justifiable apprehension caused by the practice of testing 'deterrent' weapons in the atmosphere, as well as the stockpiling of these weapons, there is also a fear of more generalized and insidious pollution due to the dumping of diluted effluents contaminated with various radionuclides originating from nuclear reactors and especially from reprocessing plants of spent nuclear fuel needed for any development of atomic energy.

The production of radioactive waste, still limited at present, is going to grow at an alarming pace from now until the end of this century. According to the OECD (Organization for European Cooperation and Development), the production of ^{137}Cs will have reached 5000×10^6 curies by 1980 and 35×10^9 curies by the year 2000. The amount of unreclaimable plutonium produced was 200 kg in 1980 and would be approximately 10 tonnes by the year 2000.

If the USA is to provide for all its electricity needs with light-water nuclear power-reactors, by the beginning of the next century it will have to deal with an annual output of nuclear waste equal to the waste from 8 million Hiroshima bombs. As an example, a power plant of the PWR (pressurized-water-reactor) type of 1000 megawatts is capable of producing annually 0.6 million curies of ^{85}Kr, 117 million curies of radiostrontium, 114 million curies of radioiodine, 154 million curies of ^{137}Cs, etc. In all, 30 tonnes of enriched uranium are needed every year to fuel such a power station, generating an equivalent mass of radioactive fuel contaminated by these various products, which must then be reprocessed.

It is obvious that the handling and administration of products contaminated by dangerous radioelements involve risks which certain do not appear to be negligible in spite of soothing statements from the nuclear power industry. Potential dangers from ionizing radiation, particularly their mutagenic and carcinogenic action, make it imperative that their reputed permissible concentration levels — when they exist — should be fixed very strictly indeed and not revised by increasing them as some people suggest.

Problems Associated with Energy

Among the problems associated with energy production and utilization, there is one worth mentioning, of a physical nature — that of thermal pollution. All machines used by man are characterized by a thermodynamic output of little more than 40% at best. Consequently, when man burns a given mass of fossil fuels or fissile material, 60% of the potential energy is lost into the environment in the form of unusable low calories. This question is particularly important regarding either classic thermal power-plants or nuclear power-plants producing vast amounts of energy in a restricted area. Cooling down a 1000 MW power-station requires the equivalent of the entire flow of water from the Seine at low water. This causes a rise in the temperature of fluvial or coastal waters which can be catastrophic for limnic and marine living organisms. On the other hand, if aircooled systems (condensation towers) are used, various unpleasant local climatic modifications can result.

2. The modern chemical industry, source of various pollutants

The Evolution of Chemical Production

The extraordinary expansion in the chemical industry over recent decades has been accompanied by emission into the biosphere of countless mineral or organic substances which are often highly toxic. Metallurgy and electronics require more and more metals and metalloids of an 'exotic' nature which are only found as traces or not even found at all among the usual constituents of living matter: mercury, cadmium, niobium, arsenic, antimony, vanadium, selenium, etc., which nowadays are used extensively in various industrial fields. As for organic chemistry, it is constantly putting into circulation an increasing number of artificial compounds. Every year more than 50 000 new molecules are synthesized and 500 new substances are commercialized on a vast scale before their toxicological properties have been studied sufficiently to guarantee the safety of their usage. The American Chemical Society estimates that there are 63 000 different chemical compounds in use at the present time.

In France, industry produces 30 million tonnes of various waste per year. In the USA, where the problems of pollution are now reaching catastrophic proportions without equal elsewhere in the world except for Japan, refuse amounts annually to 158 million tonnes of solid matter, consisting in 1978 for instance, of 59 billion empty tin cans, 32 billion bottles and barrels and 80 billion metallic and plastic wrappers and packets (the 10th Annual Report of the Council on Environmental Quality, Washington, 1979).

Polluting Agents

It is not possible to supply here an exhaustive list of the numerous organic compounds, rarely inoffensive, discharged as waste by modern industrial activity: aldehydes, phenols, fluorides, various amines, detergents, etc., which are released into the natural environment and are found in the air and waters, all of them contributing to the contamination of different ecosystems.

One particularly disturbing trend is the dispersal of various plastic materials into the environment: polythene, polyvinyl chloride, polyurethane, polystyrene, etc. It should not be forgotten that these substances, apart from possibly unpleasant traces of monomer, also contain various stabilizers, polymerizers and plasticizing agents whose toxicity has not yet been fully measured. Incomplete combustion of plastic materials, and their discharge into oceans and continental waters seem to play a significant role in the contamination of the environment by polychlorinated biphenyls (PCBs), similar substances to DDT, and by cadmium, a very toxic and carcinogenic metal used amongst various other applications as a stabilizer for certain synthetic polymers.

Global Dispersion of Certain Toxic Substances

Another important aspect of pollution of the biosphere by the chemical industry is that of the extent of areas exposed to the countless toxic substances produced by human activities. Until recently, the affected areas were localized around urban and industrial zones. But since the end of the Second World War, contamination of the natural environment by the products of modern technology is reaching more and more remote regions, showing that the menace is now on a global scale.

Even if public opinion has long been informed of the global dispersion of radioactive fallout caused by experiments with so-called 'deterrent' missiles, it is very often unaware that the same phenomenon is being caused by a large number of toxic mineral or organic substances. People have been far less concerned, until fairly recently, about contamination by synthetic chemical products of the whole of the ecosphere, including the global ocean. For instance, fragments of plastic have been found drifting in the Sargasso Sea. The atmosphere and hydrosphere are gradually being poisoned by persistent and highly toxic compounds such as hexachlorobenzene (HCB), a substance with many industrial uses, or the polychlorinated biphenyls (PCBs) mentioned earlier. Traces of these organochlorine compounds have been discovered in mammals of the Great Canadian North, in pelagic fish and even in Antarctic animals. The same is true for other substances such as mercury, used extensively in industry particularly as a catalyst and which, like HCB and PCBs, persists in all ecosystems and subtly contaminates all trophic systems.

The world's oceans occupy the final compartments, the ultimate accumulation zone for all the toxic residues produced by modern technology. It, therefore, seems incredible that modern society perseveres in treating the oceans as both a dumping place and food reservoir at the same time—two incompatible functions.

3. Modern agriculture

Causes

(a) Fertilizers—Agriculture is another important source of pollution. The extensive use of chemical fertilizers and systematic application of pesticides have caused very significant and even spectacular rises in the agricultural yields from developed countries. Unfortunately, however, this improvement in productivity has been accompanied by many undesirable effects and problems linked to contamination of the biosphere by these substances.

A determining factor in higher yields has been the extended use of mineral fertilization provided by nitrogen compounds, phosphates and potassium salts. The global use of fertilizers, which was only about 7 million tonnes in 1945, increased to over 53 million tonnes in 1968 and 108 million tonnes in 1980, excluding China.

(b) Pesticides—In the same way, the use of pesticides has seen a considerable increase, not only in the developed countries and tropical areas of cultivation for export, but also throughout the Third World, where the would-be 'green revolution' has meant a rise in the need for courses of antiparasitic treatment as new strains of crop have been used which are not as resistant as the original native crops to the different crop pests.

American production of pesticides has grown from 45 000 tonnes in 1946 to 1 million tonnes in 1980 (pure active material). About 3 million tonnes of DDT have been dispersed into the biosphere since the discovery of this insecticide. It is estimated that at least a quarter of that tonnage is at present stored in the hydrosphere and that it will remain there for several decades, even if the substance were to be totally banned world-wide, which is far from the case at the moment.

The mass of pesticides used in modern agriculture is, therefore, extremely important if we consider the incredible toxicity or persistence (sometimes both) of most of these compounds, which generally have an intense biocidal activity. The arrival of pesticides in trophic systems needs no proof. In the final analysis it concerns man himself, who is particularly exposed in his position at the top of the ecological pyramid.

(c)   Pollution of Foodstuffs

The contamination of human food is a particularly urgent environmental problem at the present time. It is of course not true that pesticides are the only chemical substances which might pollute foodstuffs. The use of antibiotics, sulphamides and even hormones in zootechnics can also lead to a disturbing contamination of our food chains. What should we think then of the voluntary use of food additives which pollute our food with colourings, flavours, stabilizers, emulsifiers, etc., and which, to put it mildly, are not particularly favourable to consumer health? There is no single dietary consideration to justify their use and they should be formally banned.

Pollution and the Global Ecological Balance

The problem of pollution is many-sided. It concerns man directly through contamination of what he inhales and ingests. Just as important, however, is the scale of both foreseeable and unexpected ecological effects. The large number and increasing diversity of the pollutants released into the environment by modern technology pose questions of their global effects on the biosphere as a whole.

If a number of contaminants are introduced simultaneously into a given ecosystem, it would probably cause eventual synergistic effects at the ecological level. The compounds released by man into the biosphere are no doubt carefully tested, generally on a few types of organisms taken in isolation. But we should be aware of the limits of these methods. In reality, various polluting agents are released and meet simultaneously, mixed together in a given biotope and affect extremely complex biocoenoses. It is virtually impossible to simulate in the laboratory either an ecosystem or even a fragment of an ecosystem.

There is a tendency to forget that toxic substances dispersed in the biosphere do not just react on a few particularly sensitive species but on a whole range of communities. Also, certain abiotic or biotic ecological factors can enhance the toxicity of substances thought to be relatively harmless, either by favouring their dispersion or by causing them to undergo chemical modifications.

Lastly, some pollution can interfere with biogeochemical phenomena and disrupt the functioning of the biosphere. Witness the recent hypothesis which claimed that pollution of the oceans by petroleum could modify the water cycle and, therefore, the rainfall regime. In fact petroleum, which has surface-active properties, limits the formation of marine aerosols. The salts released into the atmosphere by this process play an essential role in forming the condensation centres from which atmospheric water-vapour condenses.

From this it can be seen that the consequences of pollution go far beyond the effects on a given species and feature on a synecological scale.

C—THE CLASSIFICATION OF POLLUTANTS

Any logical presentation of a study of the effects of pollutants requires them to be classified. This is in fact a very difficult task as many factors have to be taken into account simultaneously.

Pollutants can of course be grouped according to their nature: physical, chemical, biological, etc., or in an ecological way either by studying their effects through levels of increasing complexity: species, population, community, or by studying the environment in which they are released and exercise their toxic potential. Finally, the classification could be done on a purely toxicological level and consider the way in which pollutants enter organisms: inhalation, ingestion, contact, etc.

Table 2.2 Classification of pollutants

Nature of the pollutants	Atmosphere	Ecosystems		
		Continental	Limnic	Marine
(1) *Physical pollutants*				
Ionizing radiation	+	+	+	+
Thermal pollution			+	+
(2) *Chemical pollutants*				
Hydrocarbons and their products from combustion	+	+	+	+
Plastic materials	+	+	+	+
Pesticides		+	+	+
Detergents			+	+
Various synthetic organic compounds	+	+	+	+
Sulphur derivatives	+	+	+	
Nitrates		+	+	+
Phosphates		+	+	+
Heavy metals	+	+	+	+
Fluorides	+	+		
Mineral particles (aerosols)	+	+		
(3) *Biological pollutants*				
Dead organic matter			+	+
Pathogenic micro-organisms	+	+	+	+

So far, none of these methods taken individually would be entirely satisfactory, as the same given substance can have many modes of action. When sulphur dioxide is released into the air, for example, it will be carried into continental waters which it then acidifies; it will also be absorbed by plants and expelled by their roots in the form of sulphate which then passes into the soil. It acts equally on plants and animals. It can be inhaled as a pollutant in the air or ingested when used as a conserving agent by the food industry, etc.

Table 2.2 shows the classification which we adopt here. It tries to achieve a compromise between the possible different criteria without masking the essentially artificial nature of such an exercise.

D — DISPERSION AND CIRCULATION MECHANISMS OF POLLUTANTS

Modern technocratic thinking commits two fundamental errors when it comes to the problems of pollution. Firstly, it considers that toxic effluents will endanger only the immediate area round the point of emission. Secondly, it claims that toxic substances will be quickly diluted in the air, soil and waters, thus diminishing their concentration to below the danger levels fixed by experts. These two assertions are always associated and considered to be complementary.

However, experience too often invalidates this simplistic conception, which ignores the complexity of biogeochemical mechanisms and of other fundamental ecological process characterizing the biosphere. The emission of pollutants into the environment is a complex phenomenon and cannot be limited to the deceptively fixed images of a plume of smoke rising from a chimney or a wastepipe spilling out its effluents into the sea. In almost all cases, substances released into the ecosphere are going to be carried a very long way from their source. Atmospheric and hydrological circulation systems will then disperse them progressively throughout the ecosphere.

1. Atmospheric transport of pollutants

The Entry of Pollutants into the Atmosphere

Atmospheric movement plays a fundamental role in the dispersion of pollutants and their distribution over different biotopes. Any organic or mineral compound, even if it is a solid, can theoretically be carried by the air. For gases, entry into the atmosphere is direct, for liquids with a weak vapour pressure it is in the form of aerosols, and for non-sublimable solids it takes the form of fine particles.

Some of the contaminants released by man into the atmosphere are natural constituents of it. Sulphur dioxide, carbon dioxide, nitrogen oxides and even mercury add to the normal quantities present in the atmosphere which come from various biogeochemical processes, and therefore natural phenomena, such as volcanism.

Other pollutants, such as radionuclides, pesticides, plasticizers, etc., are exclusively of technological origin.

The General Laws of Atmospheric Circulation

It is essential to know the general laws controlling the movement of air masses in the troposphere and stratosphere in order to understand the mechanisms governing contamination of the biosphere. These laws have been established following many investigations over the last few decades using aircraft, high performance sounding balloons and more recently with meteorological satellites.

(a) Horizontal movements—The direction and speed of stratospheric and tropospheric currents are now known with great precision. It has even been possible to show the existence of a dominant westerly wind which blows in the area of the tropopause (boundary between the troposphere, which is the lowest layer of the atmosphere, and the stratosphere; the tropopause is characterized by a rapid fall in temperatures) in the northern hemisphere. Its speed, an average of 35 m/s, means that it can carry any substance reaching this level round the world in just 12 days. This explains the rapidity with which particles from a volcanic eruption or nuclear explosion are dispersed throughout the Earth's atmosphere.

(b) Vertical movements—Combined with the horizontal currents are the vertical movements of air masses which give atmospheric circulation from north to south. The combination of west–east winds with an upward drift in the lower latitudes creates a type of atmospheric circulation called the Hadley cell. This enables the exchange of air masses between the equatorial regions of the two hemispheres to take place in the troposphere.

Between the Equator and the polar regions other cells, called Ferrel cells, come into contact with the Hadley cells, assuring the transfer of polar air masses towards the tropic and tropical air masses towards the poles.

The vertical displacements of air masses are also a determining factor in the circulation and dispersion of pollutants. The existence of massive cumulonimbus thunderstorm clouds, which can rise to an altitude of 18 km in the tropics, shows the importance of these upward movements, whose speeds sometimes go above 30 m/s.

Retention Time

Although the diffusion of pollutants in the troposphere is extremely rapid, even virtually immediate, in the stratosphere the process is very slow due to the low speed of exchange between layers of air from different altitudes. Vertical movements in the stratosphere only reach a maximum speed of a few centimetres per second, which means that particles reaching this level can stay there for years. It has been possible to calculate the average time that an unsedimentable

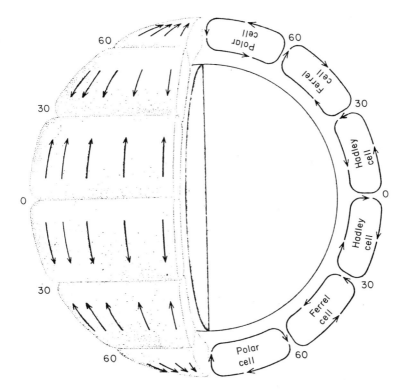

Fig. 2.2 A general model of atmospheric circulation. The combination of vertical and horizontal currents enables the transfer of equatorial air masses towards the poles and vice versa, and similarly the exchanges of air between the two hemispheres. (*From 'The energy cycle of the earth', A. M. Oort. Copyright© (1970) by Scientific American, Inc. All rights reserved.*)

particle (these particles or 'aerosols' are microscopic particles that cannot be deposited by gravity, as the Brownian movement gives them a faster acceleration rate than that of gravity because of their small mass) stays in the stratosphere as being between 2 and 3 years at an altitude of 30 km, whereas it is only 1 year in the lower stratosphere (between 15 and 18 km).

This average retention time for microscopic particles is reduced to 2 months in the tropopause layer (boundary between the troposphere and stratosphere) and down to 30 days in the mid-troposphere (about 6000 m up). Aerosols only stay for a week in the lower troposphere, at an altitude of about 3000 m (Fig. 2.4).

(a) Factors modifying retention time—Retention time in the atmosphere is longer for substances that are not affected by efficient physico-chemical mechanisms which would cause them to be removed from the atmosphere, transformed and then stored in the waters or soil. These natural processes do not work for various relatively non-reactive volatile synthetic compounds released

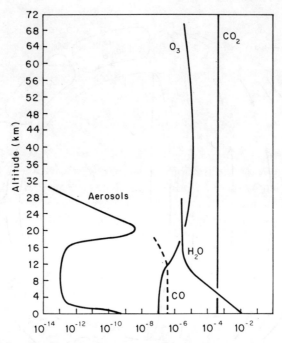

Fig. 2.3 The distribution of various natural constituents and/or man-made substances in the atmosphere as a function of altitude. Note the accumulation of unsedimentable particles in the stratosphere. (After Varney and MacCormac, 1971. *Reproduced by permission of D. Reidel Publishing Co.*)

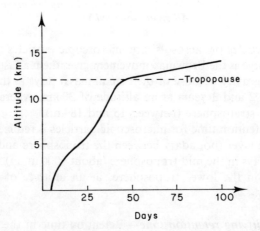

Fig. 2.4 The average lifetime of unsedimentable particles as a function of their altitude. (After Bowen *in* Lenihan and Fletcher, 1977. *Environment and Man*, Vol. 6, Blackie, Glasgow and London.)

into the air, such as freons, organic chlorofluorine compounds, and even more so for rare radioactive gases whose total absence of chemical reactively is well known.

There is, therefore, a risk of elements accumulating in the atmosphere which will remain there for an extended, though indeterminable, period of time.

(b) The case of krypton-85 — This question is particularly important with regard to the ^{85}Kr produced by nuclear reactors, whose half-life is 10.5 years. If the nuclear power programmes currently being developed in many countries of the world are brought to fruition, and in the absence of a technique for storing the ^{85}Kr produced by these reactors, it would be necessary from 2005 onwards, because of the quantities of rare radioactive gas emitted into the atmosphere, to give all personnel working in the liquid air industry protection against radiation.

2. The transfer of pollutants from the atmosphere into water and soil

Exchange Mechanisms

Fortunately, apart from a few rare exceptions, atmospheric pollutants do not remain in the air *ad infinitum*. Precipitation introduces them into the soil and/or the hydrosphere. Solid particles are carried mechanically or by dissolving, gaseous substances are dissolved in rainwater. The pollutants then circulate on the continents' surface, trickling into the soil and contaminating ground water. Processes of leaching and water erosion play an essential part in the transfer of pollutants from the soil to the hydrosphere. Finally, geochemical phenomena will eventually lead the majority of man-made pollutants sooner or later into the world's oceans, which are the ultimate receptacles for toxic agents and other pollutants produced by technological civilization.

The fundamental role of the water cycle in the transfer of pollutants was proved by studies made of radioactive fallout following the A- and H-bomb tests in the 1950s. Many later analytical studies have confirmed that the combined processes of atmospheric circulation and precipitation could transport pollutants a very long way from their point of emission.

Examples

It has been shown that the snow falling in the central regions of the Antarctic Continent is contaminated by DDT (Peterlee, 1969). Similarly, significant quantities of this persistent insecticide and a related substance, lindane, have been found in the soil of Swedish Lapland, although this area has never been treated with these pesticides (Oden *in* Singer, 1970).

Analysis of the pH of rainwater indicates that the pH level has seriously dropped as a result of the ever-increasing use of heavy fuels, rich in sulphur.

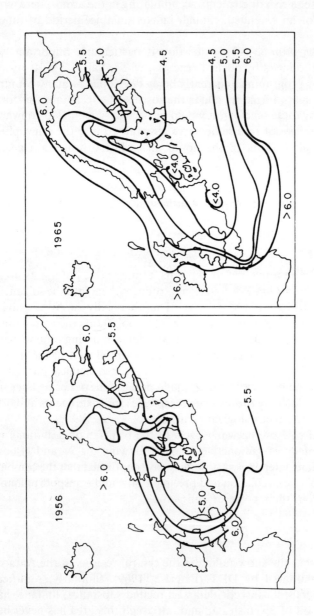

Fig. 2.5 The role of winds in pollution transport. The high acidification of rainwater falling on southern Scandinavia is now attributed to the movement of air masses polluted by sulphur from the industrial zones of Great Britain and Benelux and carried by the south-west winds. (After Oden, 1969, in Singer, 1970). Notice also the higher acidification between 1956 and 1965 due to increased consumption during this period of heavy fuels rich in sulphur

The zones where rain is most acidic are of course those where industry is concentrated, but the same phenomenon has also been found in little-urbanized areas of southern Scandinavia and in the east of the North Sea. The acidity of rain in these regions is explained by the movements of polluted air masses coming from southern Great Britain due to the effect of the dominant west winds. Therefore, various geochemical factors combine to assure the dispersal and spread of pollutants into the whole biosphere.

3. The transfer and concentration of pollutants in the biomass

The contamination of various environments by polluting agents will eventually be transferred to living beings.

The Influence of Degradability

Degradability is an important concept. Luckily, a large number of the substances released into the environment are unstable and the action of physico-chemical factors will break them down very quickly into less toxic or non-toxic derivatives. In many cases, micro-organisms — edaphic or aquatic bacteria — play an active part in this decomposition, and the substance is then called biodegradable.

Although most organic substances and even mineral elements can be converted by biogeochemical factors into less toxic or non-toxic forms, unfortunately there also exists a whole series of pollutants that are only partly biodegradable or not at all — for example, organochlorine substances and certain heavy metal derivatives.

Non-degradable Compounds

(a) The extent of contamination — The persistence of non-degradable polluting agents in ecosystems is bound to help their passage into the plant and then animal communities, that is into all the trophic systems of each biocoenosis.

The systematic study of contaminated terrestrial or marine animals with either carnivorous or ichthyophagous diets has revealed since the mid-60s the extent of pollution of the biosphere by non-biodegradable substances. These include organohalogen compounds such as DDT and various other insecticides or polychlorinated biphenyls, which are chemically related to DDT (Rizebrough *et al.*, 1968).

An analysis of various pelagic birds of the order Procellariiformes (petrels and shearwaters) who live in the most remote areas of the seas and oceans, shows the extent of contamination of the ecosphere by these substances.

Many species of birds belonging to this group are today threatened with extinction, as their exposure to such strong doses of organohalogen compounds has either partially or totally sterilized them. An example of this is the Bermuda petrel (*Pterodroma cahow*) which had become very rare due to overhunting in its colonies by man in the last century. But the later decline of this species is due to contamination of breeding individuals by the organohalogen compounds

Table 2.3 The contamination of various species of pelagic seabirds (Procellariiformes) by DDT, its metabolites and PCBs. (After Risebrough, 1971, in Hood et al., *Impingement of Man on the Ocean*, Wiley Interscience.)

Species	Place of capture (place of reproduction)	Tissues analysed	DDT and its metabolites (in ppm)	PCBs (in ppm)
Fulmarus glacialis	California (Alaska)	Whole bird	7.1	2.3
Puffinus creatopus	Mexico (Chile)	Whole bird	3.0	0.4
Puffinus griseus	California (New Zealand)	Fats	11.3 / 40.9	1.1 / 52.6
Puffinus gravis	New Brunswick (southern Atlantic)	Fats	70.9	104.3!
Pterodroma cahow	Bermudas (*id.*)	Whole bird	6.4	—
Oceanodroma leuchorhoea (Leach's petrel)	California (*id.*)	Fats *ex ovo*	953!	351!
Oceanites oceanicus (Wilson's petrel)	New Brunswick (Antarctica)	Fats	199!	697!

which pollute the ocean. Since the end of the 1950s there has been little successful reproduction and the species is declining by 3.25% per year so that it is close to complete extinction, with only about 15 breeding pairs surviving.

(b) Characteristics of their polluting effects — The extent of contamination of the biosphere by organohalogen compounds lies in their great chemical instability and consequently their weak biodegradability.

The half-life of DDT in water is estimated to be about 10 years; for dieldrin it is more than 20 years. It is, therefore, easy to understand how such substances, whose affinity for certain cell constituents is very high, are able to pass into the biomass without any difficulty. In the same way, the retention of organohalogen compounds in soil helps their uptake by living beings due to the fact that they can remain for a very long time in contaminated biotopes without undergoing any significant transformations. Table 2.4, as an example of this, shows the proportions of organochlorine insecticides to be found in an alluvial soil.

With the presence of biological concentrators in the biocoenoses, most living things can absorb the pollutants contained in the environment and concentrate them in their systems.

(c) Bioconcentration by living organisms — This phenomenon has been known for a long time due to the existence of species which are capable of accumulating natural substances in concentrations of many tens of thousands more than those usually found in soil or water. The concentration factor (Fc) is the relation between the concentration of a pollutant in an organism and the concentration in which it is found in the biotope (water, air or soil, as the case may be). For instance, the ability of algae of the genus *Fucus* or of *Laminaria* to concentrate

Table 2.4 The proportion of organo-chlorine insecticides remaining in the soil more than 14 years after their application. (After Nash and Woolson, 1967, *Science*, **157**, 925. *Copyright 1967 by the AAAS.*)

Insecticide	Percentage remaining after 14 years
Aldrin	40
Chlordane	41
Heptachlor	16
HCH	10
Toxaphene	45
	Percentage remaining after 17 years
DDT	39

iodine from sea-water, has long been used in the industrial extraction of that element.

The same phenomenon of bioconcentration is seen with the various mineral and organic substances released into the natural environment by man. For example, the plutonium in diluted waste products dumped in the ocean from treatment plants of spent nuclear fuel can be concentrated by phytoplankton at levels up to 3000 times stronger than that of plutonium in sea-water and in concentrations of up to 1200 times stronger by benthic algae.

It was in fact due to radioactive pollution that the first known cases of the bioconcentration of pollutants were discovered. From 1954, Foster and Rostenbach noticed that phosphorus 32 was found at a concentration of 1000 times greater in the phytoplankton than in the water of the Columbia river into which the effluents from the Hanford plutonium reactors were being discharged.

(d) Bioconcentrators—Plankton—The study of pollution by organohalogen compounds has also shown that in numerous biocoenoses there exist true bioconcentrators capable of literally sucking up minute traces of pollutants present in soil, water or even the atmosphere (in the case of lichens) and accumulating them in their organisms.

At the first trophic level, plants, especially those rich in lipids, such as phytoplankton in the aquatic environment and oleaginous plants on land, can store quite large amounts of very lipophilic organohalogen compounds. In carrots, whose roots are rich in terpenic derivatives, can be found a concentration of dieldrin or heptachlor equal in value to that in the soil in which they are growing: 0.67 ppm of heptachlor, an insecticide from the cyclodiene group, like dieldrin, is found in peanuts grown in soil treated with only 0.14 ppm of the insecticide. Numerous analyses have shown that these substances can be found in the parts of green plants above ground, having travelled upwards from the roots.

Phytoplankton have an astonishing capacity for accumulating organohalogen compounds. Although polychlorinated biphenyls rarely exceed a concentration of 0.1 ppb in the surface waters of the North Atlantic, they occur at a concentration of 200 ppb in the phytoplankton gathered from those waters. The record concentration level of PCBs is probably held by the marine phytoplankton growing in the Gulf of St Lawrence. Even though the PCBs only exist at lower concentrations than detection levels in this zone, up to 3050 ppb have been found in phytoplankton samples (Ware and Addison, 1973).

The Higher Levels of Trophic Chains

Bioconcentrators are also found at the higher levels of food chains. Lumbricids oligochaeta have a considerable capacity for accumulating organochlorine insecticides present in soil. Earthworms with their detritivorous diet have to ingest daily a mass of humus equal to many times their own body weight in order to meet their nutritional needs. In this way, *Lumbricus terrestris* can accumulate DDT in their organisms reaching a concentration of several tens of times greater than that of the insecticide in the humus on which the worms feed.

Certain aquatic animals are also capable of a surprising accumulation of organohalogen compounds. Bivalve molluscs with a microphagous diet can achieve very high concentration levels. It must be remembered that an oyster weighing 20 g (soft tissue) has to filter 48 litres of water a day to satisfy its dietary needs. In this way, an American species of oyster, *Crassostrea virginica*, was found to contain an accumulation of DDT in their tissues at a concentration level 70 000 times more than that of the sea-water in which they were cultivated.

The record concentration level of an organohalogen compound was found in mussels (*Mytilus galloprovincialis*) on the French Mediterranean coast, in the Marseille region. In lamellibranches, concentration factors reaching 690 000 were shown for both DDT and PCBs (in Gerlach, 1981).

Fish can also accumulate organochlorine insecticides present in water. The uptake is paradoxically not by ingestion but through the skin. Fish have in fact numerous cutaneous mucous glands which seem to play a part in the resorption of pesticides contained in water. In addition, the rapid circulation of water through the gills to ensure sufficient oxygenation of the blood also contributes to uptake of these substances into their organism. In this way, an American minnow (*Pimephales promelas*), after a few months in water contaminated with weak traces of endrin, can concentrate the insecticide to a level 10 000 times stronger than that found in its surroundings.

4. The transfer and concentration of pollutants in the trophic chains

General

Many living beings, if not all, can accumulate in their organisms, to different degrees, any slightly or non-biodegradable substance. This results in a process of transfer and biological magnification of the pollution within the contaminated biocoenoses.

Every food chain will be the site for a concentration process of the toxic substances persisting in the biomass as one progresses up through different levels of the ecological pyramid. The amounts of substances found at the end of the food chains will be greater, with all else being equal, if the substance is more stable and the food chain longer. This explains the fact that concentration levels for the same substance are always higher in an aquatic environment than in terrestrial ecosystems.

The importance of processes of concentration in trophic chains will depend on the value of the transfer factor between different successive trophic levels.

This transfer factor, Ft, can be defined as the relation between the concentration x_{n+1} of the pollutant in an organism of the trophic level $n+1$ and the concentration x_n of the pollutant in an organism of the trophic level x_n. This will give:

$$Ft = \frac{x_{n+1}}{x_n} \quad (1)$$

It is quite clear that if $Ft > 1$ there will be a bioamplification in the given trophic system, if $Ft = 1$ there will be a simple transfer without any increase in the concentration and if $Ft < 1$ the concentration of the pollutant in predatory organisms will be weaker than in those organisms in the lower ranks of the food chain.

Miller (1978) proposed a mathematical expression making it possible to quantify the transfer processes of pollutants with bioaccumulation in the trophic systems. Assume that a predator situated at the trophic levels $n+1$ with a body weight b_1 eats a_1 g of prey from the trophic level n containing a concentration x_0 of pollutant. Assume further that the predator absorbs a fraction f_1 of the pollutant which it then excretes at a daily rate of k_1. The equilibrium concentration of the given pollutant, x_1, in the body of the predator will be given by the relation

$$x_1 = Ft_{(0,1)} \cdot x_0 \quad (2)$$

in which

$$Ft_{(0,1)} = \frac{a_1 f_1}{b_1 k_1} \quad (3)$$

If the predator is then himself prey to a higher carnivore from the trophic level $n+2$, the equilibrium concentration of the pollutant in its organism will be given by the relation

$$x_2 = Ft_{(1,2)} \cdot x_1 \quad (2')$$

where

$$Ft_{(1,2)} = \frac{a_2 \, f_2}{b_2 \, k_2} \tag{3'}$$

Finally, we will have the relation

$$x_2 = \frac{a_2 \, f_2}{b_2 \, k_2} \cdot \frac{a_1 \, f_1}{b_1 \, k_1} x_0 \tag{4}$$

This can be repeated several more times as a function of the chain's length.

As the coefficients k are always a lot less than 1 for substances that are slightly or non-biodegradable ($k = 0.01$ for a substance whose biological half-life is 70 days, for example), then the transfer factors will clearly be higher than 1 and so a biological magnification that can be manyfold will take place at each trophic level.

The expression (4) makes it easy to verify that with an average transfer factor equal to 10, a magnification to the order of 100 times for the concentration of the pollutant will take place between a herbivore and a secondary predator.

The Minamata Affair

In the case of Minamata Bay in Japan, the concentration of methyl mercury in the marine food chains was reaching a rate of 500 000 times greater than that of the substance in the waters of the Bay (Ui, 1971). In fact, in limnic or marine ecosystems there are often 5 or 6 trophic levels as opposed to 3 or 4 usually in continental ecosystems.

The Pollution of Clear Lake by DDD

Since the beginning of the 1960s there has been concrete proof of the phenomena of ecological magnification of pollution by persistent organic compounds. The famous case of Clear Lake (Hunt and Bischoff, 1960) was the first ecotoxicological demonstration of this phenomenon.

The Californian lake was treated repeatedly between 1949 and 1957 with DDD, an insecticide related to DDT, in order to destroy a little midge (*Chaoborus astictopus*) which was annoying bathers, even though it does not bite.

Consequently, the DDD accumulated in the limnic food chain:

```
          water→phytoplankton→zooplankton→microphagous fish→
Trophic
levels       I              II            III

          macrophagous fish→fish-eating birds
                IV                V
```

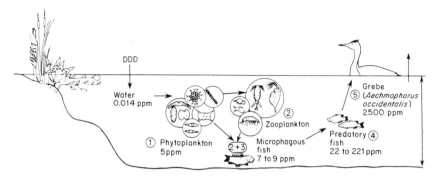

Fig. 2.6 An example of the transfer and concentration of a pollutant in a food chain: the case of Clear Lake, in California, contaminated by DDD, an organochlorine insecticide. (After numerical data from Hunt and Bischoff, 1960.)

Table 2.5 The concentration of DDD in the trophic chains of Clear Lake. (After Hunt and Bischoff, 1960)

Element or species	Trophic level	Concentration (in ppm)
Water	0	0.014
Phytoplankton	I	5
Plankton-eating fish	II & III	7 to 9
Predatory fish	III & IV	
Micropterus salmoïdes		22 to 25
Ameirus catus		22 to 221
Fish-eating birds (grebes)	V	2,500 (in the fats)

Table 2.5 shows the concentrations of DDD sampled at the end of the 1950s from the biomass of Clear Lake at the different trophic levels.

The amount of DDD sampled from the fats of the grebes (*Aechmophorus occidentalis*) reached 2500 ppm, which is a concentration coefficient of 178 500 compared to the waters of the lake. As a result, by the end of the 1950s only about thirty breeding pairs of grebes had survived on Clear Lake, and most of them were sterile, whereas the original population had been more than 3000 birds.

There are many examples concerning organochlorine insecticides which show a similar capacity for concentrating these substances in continental ecosystems.

The Contamination of Agroecosystems

Study of contamination of the food chain leading from the soil to snakes in cultivated zones of the southern United States showed a process of accumulation of another insecticide, aldrin, in the biomass.

soil (plant humus) → earthworms → toads → snakes
(*Allobophora*) (*Bufo*) (*Thamnophis*)
Trophic (*caliginosa*) (*americanus*) (*sertalis*)
levels I II III IV

The soil contained an average of 0.08 ppm of aldrin, earthworms 0.56 ppm, toads 2.31 ppm and snakes 10.5 ppm, which is a concentration coefficient of more than 1000 (Korshgen, 1970).

Hexachlorophene

There have been many other examples of the concentration of organic compounds in food chains. It was shown that hexachlorophene, a bacteriostatic agent widely used in the cosmetics industry (soaps, talcs, deodorants, etc.) which in France, for instance, had toxic effects on infants a few years ago, has in fact a strong capacity for bioconcentration in trophic chains. Samples were taken of 14 to 16 ppb (parts per billion) in the effluents from a purifying plant in a large American city, 335 ppb was found in tubeworms, 1338 ppb in crayfish and 27 800 ppb in aquatic insects. The substance could then potentially be present in freshwater fish which would feed on these different contaminated invertebrates (Sims and Pfaender, 1975).

Contamination by Radionuclides

Study of the contamination of food chains by radionuclides is also an excellent illustration of the processes of concentration of pollutants in the biomass.

Because of this, radiobiologists have been able to stress the high risks of contamination to which northern populations living in the higher latitudes are exposed from atmospheric atomic bomb tests.

Miettinen (1969) studied contamination of the Lapps by ^{90}Sr and ^{137}Cs. The Lapps are exposed to fallout of these dangerous radionuclides (strontium affects the bones, being related to calcium, and caesium affects the muscles, being related to potassium) because they are at the end of the following food chain:

soil → lichen → reindeer ⟶ milk ⟶ Lapp
(*Cladonia*) ⟶ meat ⟶

autotrophic herbivore carnivore
plant
Trophic levels I II III

Miettinen noticed a high coefficient of accumulation of the radioelements carried by the atmosphere to the lichens. More recently, Swedish scientists also working in Lapland have shown that these same cryptogams accumulate plutonium, an element that is very difficult to assimilate, in very high concentrations

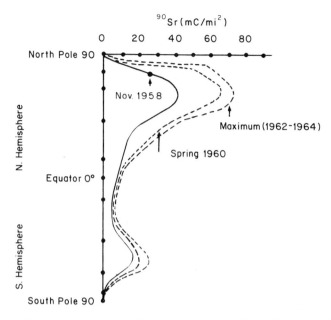

Fig. 2.7 The distribution of radioactive fallout as a function of latitude following the atomic bomb tests in the northern hemisphere. In addition to the exchange of air masses between the two hemispheres, this diagram shows that the maximum fallout was at the polar circles (66° North and South). This explains in part the high level of contamination in the food chain of the Lapps (*in* Ramade, 1982. *Reproduced by permission of Elsevier Science Publishers.*)

(from 3220 to 4290 times higher than the amounts contained in the upper soil layers).

This exceptional capacity for bioconcentration comes from the special physiology of these plants and also from the nature of the soil in the Lapland tundra. Northern soils are in fact very poor in nutritional mineral elements. They, therefore, rapidly mobilize the strontium and caesium which are then very quickly absorbed by the plants. In this way, the amount of these elements in the lichens is many thousands of times greater than that found in the tundra soil. A further concentration then takes place in the bodies of the reindeer which feed on the lichens and the Lapps then become contaminated by drinking the milk and eating the meat of the reindeer. As a result, in the middle of the 1960s the Lapps were exposed to an average dose of radiation 55 times greater than that experienced by the inhabitants of Helsinki.

5. Conclusion

All of these different investigations show how impossible it is to consider environmental pollution as a process that is easy to pin down. It is wrong to rely on dilution to spread a pollutant evenly through the soil, waters or atmosphere. Distribution is never uniform between different environments nor within a particular environment, even if transfer takes place over large distances. Certain ecosystems appear to be more threatened than others (tundra from radioactive fallout, estuaries from polluted river water). In addition, there are the processes of bioconcentration of the substances released by man into the natural environment. What happens is in effect a focusing of the pollution within food chains which transfers the toxic substances dispersed in the ecosphere towards the species situated at the highest trophic levels (primary carnivores and super predators).

If the principle of constructing ecological pyramids is applied to persistent pollutants, representing the average strength of their concentration at each trophic level, it becomes evident that this pyramid of concentrations is the complete reverse of the biomass pyramid.

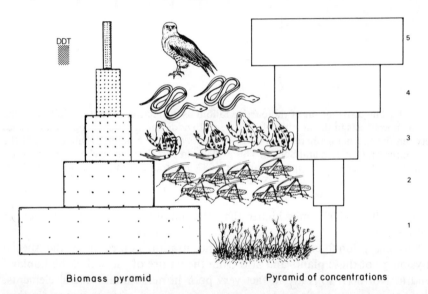

Fig. 2.8 The reverse pattern of ecological pyramids: the biomass pyramid on the left and the one of pollutant concentrations, in log units, on the right (the lesser or greater density of spots shown in the biomass pyramid is in relation to the concentration of DDT in the different trophic levels)

So, due to the process of food chains, man, who is at the top of every ecological pyramid, finds himself exposed to a sort of 'boomerang effect' from pollution.

As a result, the norms established on the simplistic criteria of inert dilution, following the effects of physiochemical factors as defined by most present-day regulations, are clearly in contempt of the most elementary ecological lessons and an unacceptable aberration.

CHAPTER THREE

Chemical Pollution

I—Pollutants affecting both continental and marine ecosystems	88
A—Organohalogen compounds	88
I—Chemical structure	89
1. Organochlorine compounds	89
2. Organofluorine compounds	92
II—Mechanisms governing the pollution of the biosphere	93
III—Demoecological and ecophysiological consequences of pollution by organohalogen compounds	103
Effects on animal populations	103
IV—The action of organohalogen compounds at the cellular level	113
V—Pathological consequences inherent to environmental pollution by organohalogen compounds	114
VI—Global effects of pollution by organohalogen compounds	115
1. Action on primary production	115
2. Effects of pollution by organohalogen compounds on biogeochemical cycles	117
B—Mercury	119
I—Principal sources of pollution by mercury	119
1. Direct causes	119
2. Indirect causes of contamination	120
II—The biogeochemical cycle of mercury	120
III—The contamination of biocoenoses and its consequences	122
1. Terrestrial ecosystems	123
2. The contamination of limnic ecosystems by mercury	125
3. Contamination of marine biocoenoses	128
IV—Physiotoxicological effects of mercury	129
Mutagenic effects	129
Mercury poisoning	130
Effects on reproductive functions	131
C—Cadmium	131
II—Pollutants of continental ecosystems	135
A—Pollutants of agroecosystems	135
I—Pollution by pesticides	135
1. Demoecological effects	136
2. Biocoenotic effects	145
II—Pollution by chemical fertilizers	146
1. The superphosphates	147
2. The nitrates	147
B—Atmospheric pollutants	150
I—Combustion products	150
1. Carbon monoxide	150
2. Hydrocarbons	152
3. Nitrogen derivatives	152
4. Sulphur dioxide (SO_2)	155
5. Ozone	167
6. Tobacco smoke	169
7. Lead	173
II—Other atmospheric pollutants	176
1. Dusts	176
2. Fluorine	179
III—Pollutants of the hydrosphere	181

I—Marine pollution	182	1. Waste from the mining industries ... 198
1. Pollution of the oceans by hydrocarbons	182	2. Pollution from lead shot ... 199
2. Pollution from detergents	193	3. Pollution by detergents ... 200
3. Pollution from solid wastes	197	4. Pollution of continental waters by various organic substances ... 201
II—Pollution of limnic ecosystems	198	

The study of the circulation of toxic substances and other pollutants in the ecosphere shows that they do not move in identical patterns between its three compartments of atmosphere, soil and water. We have already mentioned the difficulties involved in classifying chemical pollutants due to their ubiquity. However, several groups can be defined according to the nature of the ecosystems in which they are mainly found and exercise their toxic effects.

A first category of pollutants are those which affect all terrestrial, limnic and marine biocoenoses. A second category includes the pollutants found in continental ecosystems (terrestrial and/or limnic). Although these substances, like those in the first category, generally have a biogeochemical cycle which involves the continents, atmosphere and oceans, they are not yet present in sufficient concentrations in the oceans to have a damaging effect on the marine biocoenoses nor on man who consumes fish products. A third category of chemical pollutants is one whose toxic properties affect only aquatic organisms.

I—POLLUTANTS AFFECTING BOTH CONTINENTAL AND MARINE ECOSYSTEMS

A—ORGANOHALOGEN COMPOUNDS

Among the countless pollutants created by modern technology, organohalogen compounds have a particularly important place, not just because of their ecotoxicological effects but also because of their great historical interest when it comes to understanding the processes of pollution of the biosphere.

The manufacture of synthetic organohalogen compounds by the world's chemical industry, which started about 50 years ago, has steadily grown and diversified. These compounds have very varied physico-chemical properties and their application covers many spheres of activity: industrial, medical and domestic. There is no one who has not heard of DDT, initials which for lay-people have become a synonym for insecticide. This pesticide has been manufactured since 1945, with more than 100 000 tonnes produced every year world-wide. Similarly, the production of certain plastic materials, polyvinyl chlorides, is now up to more than 10 million tonnes per year. As for freons, chlorofluorine hydrocarbons, they are universally used as the cooling-medium in refrigeration plants and as gas propellants in the much-used aerosol cans which now seem to be indispensable in western households.

Whatever the merits or even considerable service given by these substances, none of them can, however, mask the important pollution problems posed by either their thoughtless or deliberate release into the natural environment.

I — Chemical Structure

At this point it is necessary to look quickly at the chemical structure of the main group of organohalogen compounds in current use.

They can be subdivided into two groups of unequal importance: organochlorines and organofluorines. In both cases they are halogen derivatives of aliphatic, aromatic and heterocyclic hydrocarbons.

1. Organochlorine compounds

These represent by far the most important group, both for the number of different substances and the tonnage produced.

Organochlorine Insecticides

These are produced nowadays in large quantities, although recently some of them have been banned from use, either partly or completely, in some industrialized countries (which does not prevent them from being exported. . . .

Historically, DDT was the first synthetic insecticide to be used on a vast scale. Its letters stand for dichloro-diphenyl-trichloroethane. It was initially employed in the fight against disease-carrying insects during the Second World War, and was later used world-wide to destroy insects damaging crops and forests. Also, for a few decades it was used extensively against mosquitoes, vectors of malaria, in vain attempts to eradicate them: attempts which were foiled by the appearance of strains of *Anopheles* resistant to the insecticide.

DDT was, therefore, dispersed into the biosphere in considerable quantities. In the absence of reliable statistics, it would not be unreasonable to evaluate the total quantity used since 1943 at more than 3 million tonnes. Virtually all the world's cultivated land, all malerial regions and many humid zones plagued by endemic parasitic diseases, several tens of millions of hectares of holarctic forests were all treated with the substance on one or several occasions, even several times a year.

Lindane, an isomer of hexachlorocyclohexane (HCH), is one of the most powerful insecticides ever discovered. As a general rule it is from 5 to 20 times more toxic to insects than DDT, according to the species. It has mainly been used against phytophagous (plant-eating) insects, but also in medical and veterinary entomology.

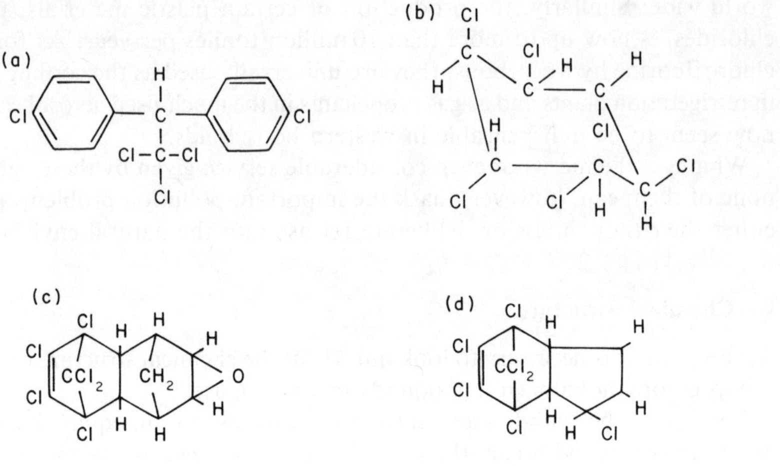

Fig. 3.1 The molecular structure of some organochlorine insecticides: (a) DDT; (b) lindane; (c) dieldrin; (d) heptachlor

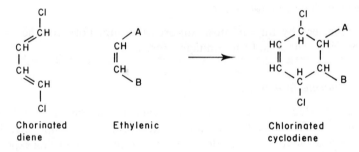

Fig. 3.2 The mechanism of the Diels-Alder reaction producing insecticides of the chlorinated cyclodiene group

Aldrin, dieldrin, heptachlor, chlordane, all chlorinated derivatives from cyclopentadiene, are not very well known, although they have sometimes been used even more extensively than DDT, for example, to combat crop pests. However, most of them are now prohibited in North America and Europe because of their dangerous ecotoxicological effects.

These heterocyclic molecules are considerably more toxic than other insecticides to both vertebrates and invertebrates. They have a great chemical stability and can persist for many decades in water, benthic silts and certain soils without undergoing any significant biodegradation.

Hexachlorobenzene (HCB), used in the past as a fungicide (seed disinfectant, fungicidal paint) is now appearing as a widespread contaminant in many environments. Although its manufacturers were very careful about its usage, the present high residual levels of the substance detected in most living organisms would suggest wide industrial use, as only this could explain such extensive contamination.

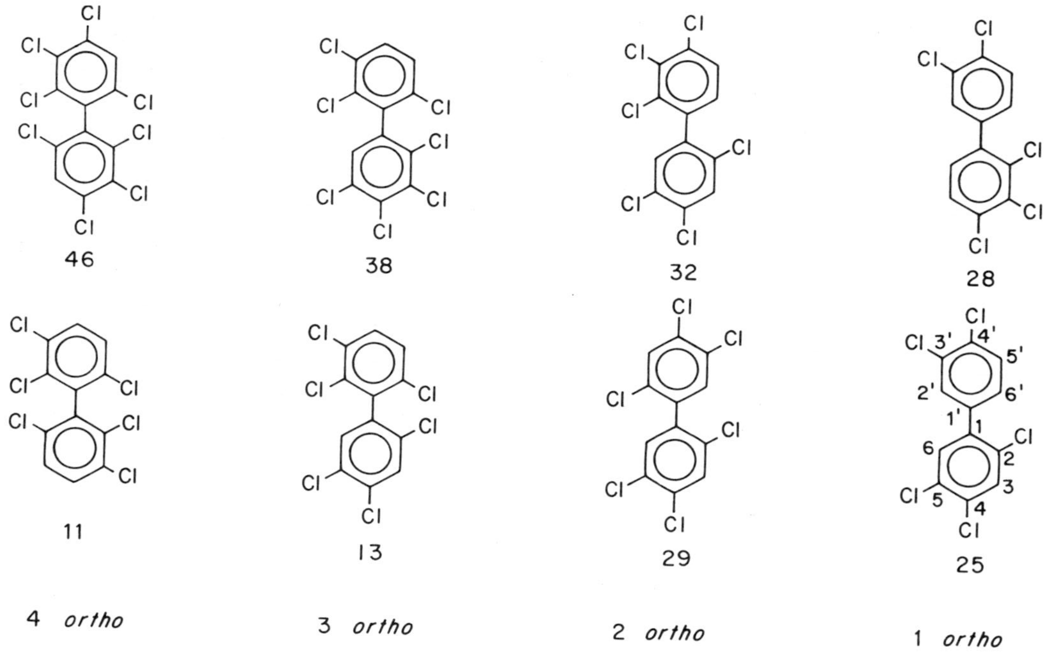

Fig. 3.3 Examples of the molecular structure of polychlorinated biphenyls (PCBs). Notice the similarity of their conformation with DDT (Fig. 3.1)

It is estimated that the present level of world production of HCB is in the order of 4000 to 5000 tonnes per annum, (Ottar, 1981). In addition, it seems that further quantities are produced as a secondary product from some industrial reactions of organic synthesis and also from the natural degradation of many polychlorinated hydrocarbons prevalent in the environment.

Polychlorinated Biphenyls (PCBs)

Their name derives from their chemical structure, which consists of a biphenyl group whose aromatic nuclei contain a variable number of chlorine atoms.

PCBs are not pure products. They are very complex mixtures of biphenyl molecules with various degrees of chlorination. In theory, their molecular framework can give rise to 189 different arrangements containing from 1 to 8 chlorine atoms. Commercial mixtures contain 4 to 8 chlorines per molecule which already allows for 102 different chemical types. They are sold under various names: 'arochlor', 'phenochlor', 'chlophen', etc., all of these preparations differing according to their physico-chemical properties and their chlorine content, varying between 32% and 62% of their weight.

PCBs were used for the first time in 1929 in the construction of electrical transformers and condensers, due to their insulating capacity and their great thermic stability. Apart from their uses in electrical engineering, they are also employed as plasticizing agents in the plastics industry, as fungicides in the paint and cardboard industries, as an 'inert' substance in the dilution of pesticides and even detergents, etc.

The total quantity of PCBs produced since their discovery is estimated at 1 million tonnes. Having reached a maximum level of 77 000 tonnes in 1970 (Risebrough *et al.*, 1976), world production of PCBs has declined over the last decade following numerous restrictions on their usage. Currently, production is about 10 000 tonnes per year (according to Ottar, 1981).

The similarity of the molecular structures of PCBs and DDT means that PCBs have very similar toxicological properties to the insecticide.

Polyvinyl Chlorides (PVC)

These are plastic substances obtained from the polymerization of vinyl chloride. Manufactured in considerable amounts, they have innumerable uses. However, apart from the physical pollution from discarded bottles and other PVC plastic packaging, there are now serious ecotoxicological questions being asked about their use in the food industry. Variable amounts of traces of the monomer, which has been proved to be a carcinogen, have in fact been found in both the solid foods and liquids kept in PVC packaging.

2. Organofluorine compounds

A large number of substances in this group are manufactured industrially.

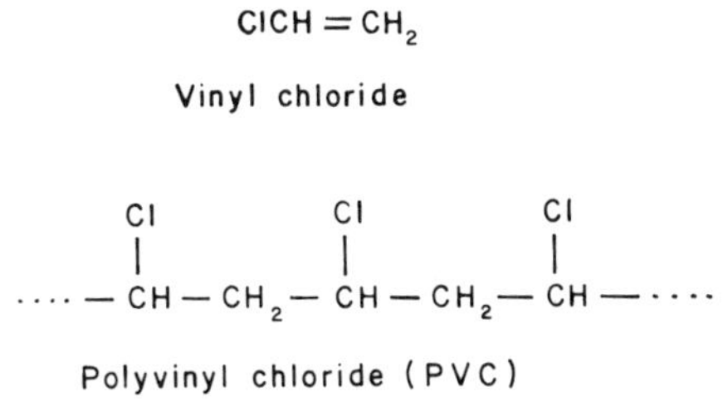

$$ClCH = CH_2$$

Vinyl chloride

$$\cdots - \underset{|}{\overset{Cl}{C}H} - CH_2 - \underset{|}{\overset{Cl}{C}H} - CH_2 - \underset{|}{\overset{Cl}{C}H} - \cdots$$

Polyvinyl chloride (PVC)

Fig. 3.4 The formula for vinyl chloride and polyvinyl chloride (=PVC), a plastic material used universally

However, as they are generally a lot more costly than organochlorine compounds they are used far less extensively.

Freons or *foranes* are chlorofluorine aliphatic hydrocarbons with a low molecular weight. Among the methane derivatives are freon 11 ($CF Cl_3$) and freon 12 ($CF_2 Cl_2$). An ethane derivative, forane 114 ($CF_2 Cl-CF_2 Cl$), is also widely used.

These gases, which are much used in the refrigeration plant industry and as propellants in aerosol cans, are remarkably inert and of a low toxicity. The growth in their use was around 15% per year in the 1960s. Their annual production world-wide was more than 1 million tonnes in 1980.

Organofluorine pesticides. Many of these are produced industrially. Dimefox, a fluorophosphorous compound with systemic properties, is absorbed through the roots and makes the plant sap toxic; it has been used against biting or suctorial insects (hop greenfly, for example). Various fluoracetic acid derivatives have similar interesting insecticidal properties. All of these different substances have an extremely high acute toxicity for vertebrates and have never shown wide possibilities for use, so nowadays they have either been virtually abandoned or else prohibited.

II—Mechanisms Governing the Pollution of the Biosphere

Sources of Pollution

These are as diverse as the uses of the substances.

One of the main pollution sources is of course the chemical factories. It is certainly not a coincidence that the most serious cases of water pollution, both limnic and marine, can be traced to chemical plants. One of the most famous cases is that of Montrose Chemicals in California. Situated near Los Angeles, this factory produced in 1970 two-thirds of the total amount of DDT manufactured in the world. From 1953 onwards, effluents from the plant were discharged into the main city sewer which in turn emptied into Santa Monica Bay, dumping some 270 kg of DDT per day into the marine environment. A change in the system of effluent disposal enabled the waste DDT to be reduced to 13 kg per day after 1971. Nevertheless, the fish and other ocean creatures were highly

...taminated in an area covering more than 10 000 km^2, going far beyond the continental shelf, as was shown by the growth in amounts of organochlorine residues found in abysso-pelagic fish from the Myctiphidae family (MacGregor, 1974). In spite of the reduction in DDT waste dumped in the Pacific after 1971, the sole (*Microstomus pacificus*) still contained 11 ppm of the insecticide in their muscles in 1975, as opposed to 17 ppm in 1971 (Young *et al.*, 1977).

In the same way, the pollution of the Rhine downstream from Rotterdam as well as the Dutch North Sea coast by dieldrin and telodrin was caused by factories producing these insecticides (Koeman *et al.*, 1967).

The problem is identical for PCBs, as both manufacturing plants and factories using them extensively play an important part in the contamination of aquatic environments. Holden (1970) was able to show, for example, that the industrial waste in the Clyde estuary was equivalent to a minimum of 1 tonne per year of PCBs. Duke *et al.*, 1971, incriminate first of all the industries concerned in the pollution by PCBs of the Florida coast.

Next as sources of pollution are the individual users of organohalogen compounds who also contribute significantly to the dispersion of these substances into the environment. A typical case is that of aerosol cans containing freons, whose usage is primarily individual. Certain aerosol deodorants contained until quite recently another organohalogen compound, hexachlorophene, a strong antiseptic that is very toxic to mammals. A famous dramatic incident involving this was the case of the Morhange talc, which led to the serious poisoning of more than 200 infants in France in the spring of 1972, due to a cosmetic preparation with a talc base which, because of a handling error, contained an excessive dose of hexachlorophene (6.3%). Thirty-four of the young children died and the others had differing degrees of damage to their central nervous system, which was partly irreversible (Martin-Boyer *et al.*, 1982).

This very stable antiseptic substance is still to be found in the liquid effluents of all large towns and cities.

The uncontrolled incineration of household and industrial waste, accompanied by incomplete combustion, is a significant contributory factor to the atmospheric pollution by PCBs and other organohalogen compounds.

In contrast to the sources of pollution just cited, which are for the most part unintentional, the use of pesticides is a case of deliberate dispersion of substances with far-reaching biocidal properties into the natural environment. Furthermore, the considerable chemical stability of organochlorine insecticides increases their pollution potential. They can persist for a long time in the environment, facilitating their transport by biogeochemical processes to areas far removed from the regions of original application.

Spraying by machines on the ground or from the air is a direct cause of atmospheric pollution. At best, little more than 50% of the active substance is immediately taken into the air at the moment of application. In addition, processes of codistillation with water vapour, even in the case of dieldrin or DDT which are molecules with weak vapour tension, carry into the air a significant fraction of the insecticide which had settled on the foliage or the soil.

Atmospheric Circulation and Contamination of Other Areas of the Biosphere

Eventually, the uptake into the atmosphere of organohalogen substances, either in gaseous form or as suspended particles carried by the winds, will result in general contamination of the troposphere. The compounds will then be brought down to the surface of the land or sea by precipitation, reaching often very remote areas.

Processes of run-off and leaching are accompanied on the other hand by accumulation in soils. But leaching can also contribute to the general pollution of continental surface waters and ground waters. Also, the rivers will carry contaminated effluents through the estuaries to the continental shelf, with the ocean being the ultimate sink, as it is for all the other waste products of human activity.

The processes of atmospheric circulation along with those of the biogeochemical cycle of water will ensure the dispersion of organohalogen compounds throughout the biosphere.

For quite a long time now there has been concrete proof of the global transport of these substances and their transfer by geochemical processes between the continents, atmosphere and hydrosphere. Since the beginning of the 1960s, samples of air taken from the lower stratosphere above the frozen north of Canada showed the presence of traces of organochlorine insecticides. Also, residues of the pesticides were found in various antarctic animals (seals and penguins in particular) which were soon attributed to an *in situ* contamination of these species, Peterlee (1969) having discovered traces of DDT in the snow falling in the central region of the Antarctic Continent.

Risebrough *et al.* (1968) gave further proof of the transport of organochlorine compounds over great distances. They found 41 ppb of DDT in the atmospheric dust falling on Barbados. This came from soil particles brought by wind erosion from the southern regions of Morocco, which had been treated with anti-acridians. The insecticide had, therefore, managed a journey of some 6000 km over the Atlantic before coming down on the island of Barbados.

Another example showing the global redistribution of organochlorine pesticides is their appearance in the soils of northern Europe that have never been treated with them. Oden (*in* Singer, 1970) discovered traces of lindane and DDT in samples of soil taken from 500 locations in Sweden, all places where the insecticides had never been used. One remarkable fact is that the concentrations of these substances increase as the samples are taken further north, echoing the distribution by latitude of radioactive fallout found in the same zone of the northern hemisphere (Fig. 3.5).

A final proof of the role of atmospheric transport in the pollution of the continents and the hydrosphere by organohalogen compounds was supplied by Ware and Addison (1973). These scientists demonstrated a positive correlation between the amount of PCB contained in marine phytoplankton gathered from the Atlantic off the Gulf of St Lawrence and the amount of precipitation in

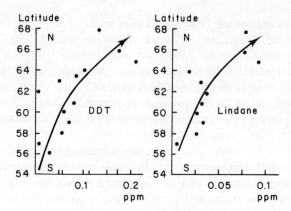

Fig. 3.5 The accumulation of DDT and lindane, in terms of the latitude, in Swedish soils which had never been treated with these substances (atmospheric transport). (Diagram by Oden, quoted by Lundholm *in* Singer, 1970, p. 201. *Reproduced by permission of D. Reidel Publishing Co.*)

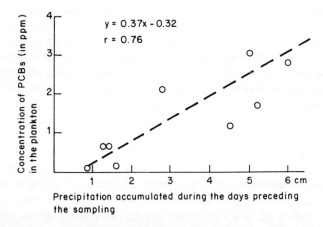

Fig. 3.6 The correlation between the amount of precipitation in the Gulf of St Lawrence between 10 and 20 days before the sampling of plankton and the amount of PCBs found in the samples (in ppm/fresh weight). There is a clear correlation between the amount of rainfall and the degree of contamination of the plankton, proving that the PCBs are being carried in the atmosphere to the Atlantic. (After Ware and Addison, 1973. *Reprinted by permission from* Nature, **246**, *5434, 519–521. Copyright© 1973, Macmillan Journals Ltd.*)

the same zone during the days preceding the collection of the samples analysed (Fig. 3.6).

Incorporation into the Biomass

The great chemical stability of organohalogen compounds helps their uptake and storage in living beings. These processes were described in Chapter 2 and will not be given here again in detail. However, it is worth repeating the fact that phytoplankton in the aquatic environment and various phanerogams on land have an extraordinary capacity for accumulating organochlorine insecticides and PCBs. For instance, dieldrin, undetectable in the waters of the North Sea, was found either in doses equal or greater than 1 ppb just off the British coast in the middle of the 1960s (Robinson, in Ramade 1982).

At the present time it seems that pollution of marine phytoplanktonic organisms by PCBs is greater than by organochlorine insecticides. Harvey, *et al.* (1974) found a ratio of PCB/DDT and metabolites equal to 30 in 53 plankton samples taken from deep-sea zones in the Atlantic between the latitudes 66°N and 35°S. Concentrations of PCB to the order of 200 ppb are usual in these remote ocean regions and these researchers often found 1 ppm in the phytoplankton, showing these organisms' large capacity for accumulating organohalogen compounds.

A large number of animal species also act as 'bioconcentrators' for PCBs and organochlorine insecticides (cf. Ch. 2, pages 78–80). Generally speaking, they are either microphagous or detritivorous animals who are able to support high concentration coefficients.

In the limnic or marine environments, apart from bivalve molluscs and other animals who filter their food, various detritivorous species such as crustaceans and annelids also have the ability to take in the organohalogen compounds present in the biotope. For instance, fiddler crabs (*Uca pugnax*), which feed on dead vegetable matter (remains of aquatic plants and associated bacteria, etc.) contained in the coastal sediments and contaminated with DDT, can very quickly accumulate residues of the substance in their muscles (Odum *et al.*, 1969).

A good correlation has been shown between the concentration factor of organohalogen compounds in the organism of molluscs and marine fish and the degree of insolubility of these compounds in water (Fig. 3.7). Thus for lamellibranchs the concentration factor for lindane, whose saturation point in water is 10 ppm, is 100 times on average, and the factor is nearly 100 000 for hexachlorobiphenyl, whose saturation point in water is 1 ppb (Ernst, 1980).

The Transfer and Accumulation of Organohalogen Compounds in the Food Chains

In Chapter 2 it was explained how most living beings, if not all, are capable of storing any chemically stable substance in their organism and that this is the case with organohalogen compounds, with a resulting process of transfer and biological magnification of the pollution within the contaminated biocoenoses. Every food chain will, therefore, be host to a process of concentration of organohalogens that increases as one goes up the trophic levels it contains.

There are now many examples of organochlorine insecticides accumulating in considerable amounts in the food chains. The case of the DDD contamination of Clear Lake has become a classic, because it was one of the first examples showing the dangerous ecological consequences of pollution.

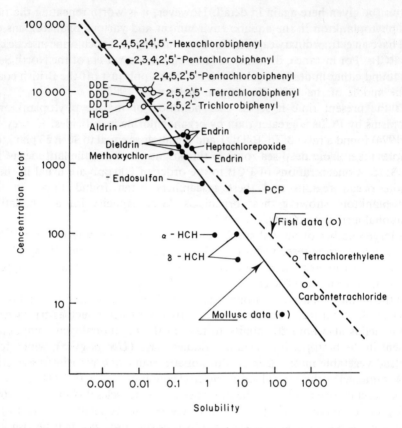

Fig. 3.7 The correlation between the concentration factor of organochlorine compounds in marine organisms and the degree of insolubility of these compounds in water. (After Ernst, 1980.)

Much experimental proof now exists for the phenomena of concentration of organohalogen insecticides in the terrestrial environment (cf. Ch. 2, page 82 ff). More recently (the end of the 1960s) various studies started to show that other substances apart from organochlorine compounds, namely PCBs, were capable of accumulating in the trophic systems, although this is not surprising given their structural relationship to DDT. Dustman et al. (1971) studied the kinetics of PCBs in the coastal biocoenoses of the eastern coast of Florida (Escambia Bay). They found the following amounts: in water up to 275 ppb, in oysters 2 to 3 ppm, in prawns (*Penaeus duorarum*) 1.5 to 2.5 ppm and in fish (*Lagodon rhomboides*) 6 to 12 ppm.

The Importance and Extent of Pollution of the Biosphere by Organohalogen Compounds

Biogeochemical processes have led to the pollution of all terrestrial, limnic and marine ecosystems by these substances. At the present time every square metre of the planet has received its molecular quantum of organochlorine insecticides and PCB (atmospheric transport).

Even in the most remote zones, where there is no human activity, the biomass is contaminated. As well as the examples quoted earlier of petrels and shearwaters, pelagic seabirds that spend the majority of their existence on the high seas (cf. Table 2.3, page 78), there are many others showing the extent to which the world's oceans are polluted. For instance, tunny-fish (*Thunnus albacares*) caught in areas off the Galapagos Islands contained 0.02 ppm of DDT, and other fish of the same species caught in pelagic zones near the Equator had 40 ppb of PCBs (compared with their wet weight).

Harvey *et al.* (1974) showed that all zooplankton and nekton living in the Atlantic between 66° North and 35° South contained higher doses of DDT and PCB than 1 ppb and often up to 1 ppm. Superpredators (*Carcharhinus longimanus*) caught in the central zones of the Atlantic had up to 13 ppm of PCBs in their hepatic fat.

The extent of pollution by organohalogen compounds is testified by the presence of organochlorine insecticides and PCB in many species of birds living in northern subarctic regions. For instance, Alaskan bald eagles (*Haliethus leucocephalus*) contain 1.65 ppm of DDT (wet weight), and peregrine falcons (*Falco peregrinus*) from the same region have 95 ppm of DDT and its metabolites (dry weight) in their pectoral muscles. Even bears and polar foxes killed on the inhospitable west coast of Greenland contain quite

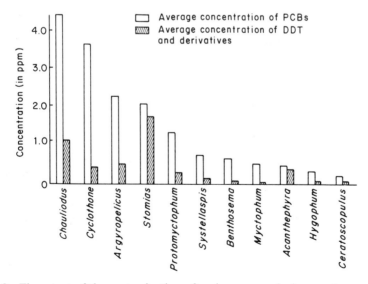

Fig. 3.8 The extent of the contamination of various mesopelagic organisms collected in the Atlantic Ocean between 66°N and 35°S. The amount of PCBs and DDT and its derivatives in the lipids is expressed in ppm (After Harvey *et al.*, 1974.)

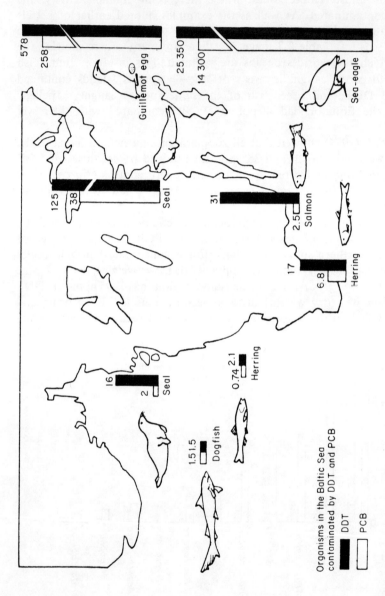

Fig. 3.9 The contamination by organohalogen compounds of various marine animals in the Baltic Sea and North Sea. Notice the considerable concentrations found in the sea-eagle and guillemots, predatory birds at the top of the ecological pyramid. The figures give the concentration in ppm of DDT and PCBs (in the lipids). (After Jensen *et al.*, 1969, but modified by Barde and Garnier, 1971. *Reprinted by permission from Nature, **223**, 5216, 753–754. Copyright©1969 Macmillan Journals Ltd.*)

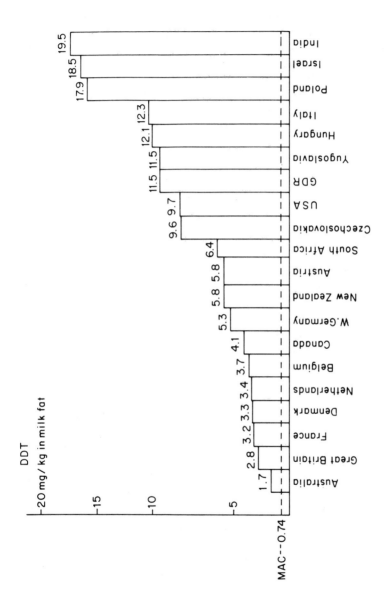

Table 3.10 The average concentration of DDT in the fats of mothers' milk. It is noticeable that it is systematically above the maximum admittable concentration (MAC) in human foodstuffs. (*In* Schüpbach, 1981.)

significant amounts of DDT, DDE, lindane, heptachlor and aldrin (Clausen et al., 1974).

Near densely populated and industrialized zones, such as on the shores of the Baltic Sea, pollution by organohalogen compounds can reach catastrophic proportions. Jensen et al (1969) found up to 36 000 ppm of DDT and its metabolites and 17 000 ppm of PCB in the lipids of sea-eagles (*Haliaëtus albicilla*) nesting on the Baltic coast.

During the 1970s, analyses of mothers' milk carried out in all industrialized countries and in certain regions of the Third World showed that the average concentration of DDT was systematically above the "tolerable level" in human foodstuffs (cf. Fig. 3.10). The record was held by the Indian with a concentration of 19.5 mg/kg, 26 times the MAC (maximum admittable concentration) (in Schüpbach, 1981).

Analyses of mothers' milk done in 1980 in the Paris region showed an alarming contamination of the milk by PCBs. In samples of milk taken on the 5th day of breast feeding from a variety of mothers there was an average of some 313 ppb of DP5 and 85 ppb of DP6 (8.77 ppm and 2.47 ppm respectively of the milk fats). On the other hand, the amount of pesticide was less (32 ppb of DDE, 7.7 ppb of HCB and 3 ppb of heptachlorepoxide), which should be related to the measures taken prohibiting these substances in France at the beginning of the 1970s (Doubilet, 1981).

Coastal communities are particularly exposed to pollution by organochlorine compounds in industrialized countries. A survey of all the main river basins in the United States, taking as its reference sample a mixture of dominant species of fish, showed that the contamination of these ecosystems was reaching disturbing levels (Veith et al., 1979). PCBs were detected in 93% of the fish analysed, 53% of them had more than 5 ppm. Only 14% of the samples had less than 2 ppm, considered to be the MAC, with concentrations varying between 0.3 ppm and 140 ppm. In addition, the fish sampled contained between 50 ppb and 4.53 ppm of DDT and its metabolites, depending on their source. Chlordane was detected in 36% of the samples and hexachlorobenzene in 14% of them.

Situated at the very top of the ecological pyramid, man cannot escape contamination from these substances. In analyses made more than 20 years ago the average American already had 12.9 ppm of DDT in his fats, the Frenchman had 5.2 ppm, the German 2.3 ppm and the Briton 2.2 ppm (Hayes and Dale, 1963). It has also been possible to show traces of PCB in adipose tissue and human serum. In 1971 it was established that 91.7% of the population of the State of Michigan contained PCB residues to a level above 1 ppm in 47.2% of the people studied (Fishbein in Kraybill et al., 1972). At the end of the 1960s, the level of contamination of mothers' milk by DDT was deemed high enough by the Public Health authorities in the State of California for them to advise mothers against breast feeding their babies. The excessive amounts of DDT observed were due to pollution of the following food chain: soil→grass→cows→milk→mothers.

III — Demoecological and Ecophysiological Consequences of Pollution by Organohalogen Compounds

Effects on animal populations

Any long-term and ecologically significant change in a population is always difficult to prove and to ascribe to precise causes. However, there is now concrete proof of the determining role played by various organochlorine compounds in the decline or conversely in the rapid multiplication of many animal populations.

Effects on Invertebrates

Many different damaging effects resulting from the dispersion of organochlorine insecticides over vast areas were very quickly discovered. They come from disrupting biological balances when a pesticide modifies the competition between species.

(a) Effects on the entomofauna — Use of DDT against *Citrus* pests in California was followed by a proliferation of citrus scale insect (*Aonidiella aurantii*) due to the earlier destruction of the predators and parasites of this species (De Bach *et al.*, in Huffaker, 1971).

There are other cases where the elimination by an insecticide of rivals for the food of a phytophagous species results in rapid multiplication of that species. At the end of the 1950s in the Cañete valley in Peru there were six new species of insects damaging cotton plants, all of them appearing as a result of daily treatments of the crop with organochlorine insecticides.

Systematic use of persistent pesticides can also favour the development of strains of insects who are resistant to them. For instance, at the beginning of the 1980s such strains had been detected in about 450 species of Arthropoda that are crop pests or that transmit serious parasitic or viral diseases (Georghiou and Mellon, 1983).

(b) Effects on limnic and marine invertebrates — In contrast, a large number of species of terrestrial, limnic or marine invertebrates are extremely sensitive to organohalogen compounds. Repeated spraying with DDT over more than 15 successive years of the coniferous forests of New Brunswick in Canada, covering areas of nearly 10 million hectares, caused a significant reduction in the numbers of aquatic invertebrates inhabiting the Miramichi and Sevogle rivers. The larvae of various freshwater insects of the orders: Ephemeroptera, Plecoptera and Trichoptera on which young salmon feed were virtually eliminated during the period following each treatment (Keenleyside, 1967). As a result, there was a decrease of 85% in the number of salmon caught in the State during the 1960s (Kerswill, 1967).

In another study it has been shown that development of the eggs and survival of the larvae of a lamellibranch molluscs are strongly affected by organochlorine

Table 3.1 Effects of various organochlorine insecticides on the development of eggs and larvae survival of lamellibranch molluscs. (After Davis and Hidu, 1969)

Insecticide	Concentration (in ppm)	Species	Percentage of development	Percentage of larvae survival
Aldrin	0.25	*Mercenaria*	90	75
	1	*mercenaria*	17	0
DDT	0.2	idem	91	88
(in suspension)	2		60	94
Dieldrin	0.025	*Crassostraea*	95	69
	0.25	*virginica*	67	58
Endrin	0.025	idem	100	79
	0.25		52	67
Toxaphene	0.25	*M. mercenaria*	89	33
	1		51	0

insecticides. Davis and Hidu (1969) looked at the effects of these substances on two species (*Mercenaria mercenaria* and *Crassostraea virginica*). They showed that development was affected even by concentrations of less than 1 ppm. (cf. Table 3.1).

Other more recent research has shown that minute concentrations of organochlorine insecticides or PCB impede the growth of molluscs. One ppb of 'Arochlor' 1254 is enough to cause a reduction of 20% in the shell growth of the oyster *Crassostraea virginica*.

Zooplankton crustaceans are hypersensitive to these substances. The transformation of nauplii into adults is inhibited in the copepod *Pseudodiaptomus cornatus* by only 0.1 ppb of DDT. No prawn (*Penaeus duorarum*) can survive 48 hours of 100 ppb of PCB ('Arochlor'). The record for hypersensitivity to PCBs is held by the amphipod *Gammarus oceanicus* whose lethal threshold after 30 days of contact with 'Arochlor' 1254 is less than 0.10 ppb.

Effects on Vertebrates

(a) Effects on fish—Pollution by organohalogen compounds is always detrimental to populations of contaminated Vertebrata. It is certainly responsible for the growing scarcity of various Salmonidae. For example, pollution of Lake Michigan by DDT and its metabolites resulted in the virtual disappearance of the Coho salmon (*Oncorhynchus kisutch*) and the Kiyi salmon (*Coregonus kiyi*) which were formerly abundant in the lake.

The eggs and fry of fish are in fact particularly sensitive to organohalogen compounds. A concentration of 5 ppm of DDT in water causes 48.3% mortality of carp embryos in the egg. This rate goes up to 93.7% with chlordane and 100% with dieldrin and endrin.

Two of the author's colleagues, Habib Boulekbache and Catherine Speiss (1974) demonstrated that 1 ppm of lindane lowered by 20% the hatching of trout eggs and caused serious morphogenetic abnormalities, particularly in the fry's caudal region. A concentration of 1 ppm is fatal to the fry of marine fish and causes an incredible reduction in the growth of the young fish. For freshwater Teleostei, no species has an LC_{50} after 48 hours higher than 20 ppb, except with lindane.

The sensitivity of fish to PCB is also very high. The LC_{100} after 45 days is around 5 ppb for *Leiostomus acanthurus*. For the rainbow trout (*Salmo gairdneri*) the LC_{50} for compounds related to PCBs, terphenyls, is 10 ppb after 48 hours of exposure.

(b) Effects on the avifauna—Many bird populations have been victims of pollution by organochlorine compounds. In the majority of cases, the virtual extinction of certain species inhabiting vast areas is not a result of acute poisoning with spectacular deaths of the adults. Rather it comes from more unobtrusive physiological disorders affecting the biotic potential of contaminated individuals.

Over the last couple of decades, ecologists have been alarmed by the growing scarcity of numerous species of fish-eating birds and birds of prey which had previously flourished. Many falcons, particularly the peregrine falcon (*Falco peregrinus*) have seen a catastrophic reduction in their numbers. At the start of the last decade the species had virtually disappeared altogether in certain areas of Scandinavia, of Great Britain and in the United States (Peakall, 1970). It had become very rare elsewhere (e.g., France). The osprey (*Pandion haliaëtus*) had severely declined in number on the east coast of the US. The Connecticut colony, which had about 150 breeding pairs in 1952 had dropped to 10 birds by 1970 with an average of 0.23 fertile young per nest, far below the number required to maintain the population (Wiemayer *et al.*, 1975).

Presently, the previously dwindling populations of raptors birds are showing significant signs of recovery, as a consequence of the widespread ban of organochlorine insecticide at the beginning of the 1970s.

In Western Europe, for example, peregrine falcon, golden eagle and other threatened species are now increasing in numbers. Ratcliffe (1980) pointed out that eggshell thickness index has displayed an upward trend since the mid-1970s.

The main cause of the decline of fish-eating birds is contamination of their prey by insecticides and PCBs. Another example is that there were only about 1500 breeding pairs of American bald eagles surviving in the whole of the US (of which it is the national emblem) at the end of the 1970s. Despite an improvement of its reproduction since the ban of DDT in 1972, this species still had pesticide and eggshell thinning problems at the beginning of the 1980s (Scott and Eschmeyer, 1980). Another fish-eating eagle, the sea-eagle (*Haliaëtus albicilla*) has virtually disappeared from Europe, its major nesting sites round the Baltic Sea being highly contaminated by organochlorine compounds (cf. fig. 3.9). In Great Britain and Scandinavia, the sparrow-hawk (*Accipiter nisus*), golden eagle and various Strigiformes have also become very rare.

It has also been possible to find a link between the significant decline of many colonies of waterbirds and shorebirds, which were flourishing in the recent past, and a reduction in their breeding success due to organochlorine pollution. Koeman *et al.* (1967) showed for example that the disastrous decline in the colonies of sandwich tern (*Sterna sandvicensis*) on the Dutch coast (from 50 000 breeding pairs in 1952 to only 150 in 1965) resulted from a reduction in the biotic potential of this species which had been highly contaminated by dieldrin and telodrin residues present in invertebrates and fish off the North Sea coast.

Similarly, from 1964 onwards considerable doses of DDT residues and other organochlorine insecticides were found in 17 species of British sea-birds (Moore and Tatton, 1965).

The most spectacular case of the depletion of a species of bird that can be attributed without question to pollution by organochlorine compounds is that of the colonies of brown pelican (*Pelecanus occidentalis*) in the United States. Both the eastern sub-species (*P.O. carolinensis*) and the Californian sub-species (*P.O. Californicus*) declined drastically between 1960 and 1970. The colonies in Louisiana have almost disappeared. Those of the Anacapa islands (southern California) dropped from 3000 breeding pairs in 1960 to only 300 in 1969. Among the remaining 300 there were many unsuccessful attempts at hatching their eggs: 1200 eggs were laid in 1969 but only 5 viable chicks were born.

Effects on the Fertility of Birds

The first studies of the damaging potential of organochlorine insecticides were made in the laboratory by de Witt (1955) and by Genelly and Rudd (1956). In the second study, now a classic, the researchers showed that dieldrin, DDT and toxaphene, added in amounts from 25 to 100 ppm to the feed of pheasants caused a clear decline in the number of eggs laid and also affected the viability of the chicks.

Since then, much research has contributed to the analysis of the physiological processes by which organohalogen compounds lessen the biotic potential of species of birds. Numerous factors combine to affect fertility. Some of the most important include the date of the first laying, a reduction in the number of eggs laid, sometimes even the total sterility of the female and her inability to lay a replacement clutch of eggs.

It was only at the end of the 1960s that concrete proof was established of an essential process in the decline of various species of birds due to pollution from organochlorine compounds: thinning of the eggshells because of abnormal calcification. Ratcliffe (1967, 1970) was the first to draw attention to this phenomenon. He was intrigued by the disturbing correlation in Great Britain between the amount of organochlorine insecticides found in the eggs of certain birds of prey, the abnormal fragility of the eggshell and the frequent failure to reproduce observed in these species. For the peregrine falcon, Ratcliffe noticed that there had only been 3 known cases of eggs damaged in the nest in 109 eyries studied between 1904 and 1950, whereas 47 damaged eggs had been found in 168 eyries visited between 1952 and 1961. In a comparative study of samples of eggs conserved in various British museums with recent samples from the nests

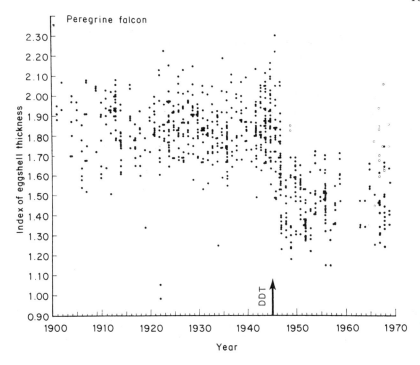

Fig. 3.11 Effects of environment pollution by DDT in Great Britain on the eggshell thickness of peregrine falcon eggs. The biometric study of eggs conserved in museums since the beginning of the century compared with more recent specimens shows a noticeable reduction in the index of eggshell thickness after 1946, the date when this insecticide was introduced on a vast scale. (After Ratcliffe, 1970. *Reproduced by permission of Blackwell Scientific Publications Ltd.*) ● eggshells from England, ○ eggshells from the Scottish Highlands

of different species of birds of prey, Ratcliffe was able to show that the index of eggshell thickness (ratio of eggshell weight to the axial length of the egg) decreased suddenly between 1945 and 1946, years which saw the introduction of organochlorine insecticides on a large scale.

The conclusions taken from this research have been largely corroborated by later studies. One of these, made by Anderson *et al* (1972), who measured 23 658 eggs from 25 different species of birds sampled from the beginning of this century to the date of their study, confirmed the significant decrease in eggshell thickness from 1946 onwards for bird populations in vast areas of the northern hemisphere.

Any differences observed can be attributed to the relative levels of contamination from region to region, diet and specific variations in sensitivity to organochlorine compounds. However, a decrease in the index of eggshell thickness has now been proved for at least 11 families of birds: Gaviidae, Procellariidae, Pelecanidae, Phalacrocoracidae, Ardeidae, Accipitridae, Falconidae, Laridae, Scolopacidae, Strigidae, Corvidae.

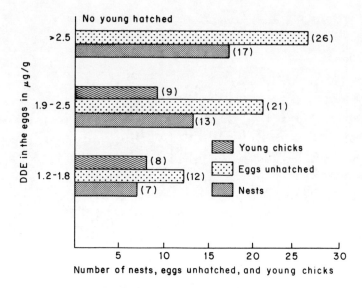

Fig. 3.12 Correlation between the amount of DDE in the eggs and the breeding success of the brown pelicans of South Carolina. Notice the total lack of breeding success in nests where the eggs contain more than 2.5 μg/g of DDE, the main metabolite of DDT found in these birds. (After Blus *et al.*, Reproduced by permission of Pesticides Monitoring Journal.)

The record for eggshell thinning is unquestionably held by the pelicans from the Anacapa islands (California). Two-thirds of the nests of this colony had no eggs in 1969 and the remaining one-third had only broken or dehydrated eggs (*in* Stickel, 1973). Some of the eggshells, almost totally lacking in calcium, had been reduced to a thin chorionic membrane which broke open when the scientists tried to take them for testing. Such eggs contained up to 2500 ppm of DDE, the main DDT metabolite, in their fats. In fact, of the 300 breeding pairs making up the colony in that year only 12 had apparently normal clutches of eggs.

The brown pelicans of the eastern coast of the United States have also shown a clear increase in the fragility of their eggs over recent decades. In the colonies of South Carolina, where the numbers of birds went down from 5000 breeding pairs in 1960 to 1200 in 1969, the index of eggshell thickness decreased by 17% over the same period. A comparison of the residual amounts of DDT, DDE, DDD, dieldrin and PCB made between eggs from normal clutches in the colony and those from nests where the eggs had failed to hatch, showed significant differences (Blus *et al.*, 1974) (cf. Fig. 3.12 and Table 3.2). A strong positive correlation was, therefore, established between the infertile clutches and the amount of DDE and dieldrin in the eggs.

The role of DDT and DDE in inducing the phenomenon of eggshell thinning was proved experimentally at the end of the 1960s. Heath *et al.* (1969) observed that the addition of 40 ppm of DDE to the food of female mallard ducks

Table 3.2 Contamination of the eggs of Carolina pelicans by organochlorine compounds. (After Blus *et al.*, 1974. *Reproduced by permission of Elsevier Applied Science Publishers Ltd.*)

Nest	Geometric average expressed in µg of residues per gram in relation to the gross weight of the egg				
	DDE	Dieldrin	PCB	DDT	DDD
With normal clutch	1.77	0.30	5.50	0.11	0.30
With infertile clutch	3.23*	0.49♦	7.94	0.17	0.46♦

♦ significant threshold at p 0.025
* significant threshold at p 0.005

(*Anas platyrhynchos*) caused a reduction of 13.5% in the eggshell thickness as well as a decrease of 34% in the number of chicks eventually hatching.

Other research carried out on females of the same species of duck, this time given 1.64 and 10 ppm of dieldrin in their food, showed that even the weakest dose of this insecticide was likely to cause a significant reduction in the eggshell thickness (Lehner and Egbert, 1969).

Similar experiments conducted on falcons (Porter and Wiemeyer, 1969) and on North American nocturnal birds of prey have also shown that DDE causes considerable eggshell thinning in these species.

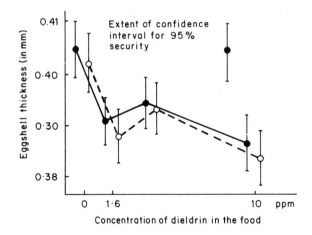

Fig. 3.13 Influence of contamination by dieldrin in the food of mallard ducks (*Anas platyrhynchos*) on eggshell thickness. ● = equatorial thickness, ○ = polar thickness (After Lehner and Egbert, 1969. *Reprinted by permission from* Nature, **224**, 5225, 1218–1219. Copyright©1969 Macmillan Journals Ltd.)

PCBs seem to be less dangerous to eggshell formation in doves, mallards and falcons, although this could be due to various differences linked to the nature of PCBs. Some Arochlors are inactive, whereas Arochlor 1254 administered in high doses (100 ppm) in chicken feed has caused both a reduction in the number of eggs laid and in the shell thickness.

Perturbation of the Endocrinal Balance in Species of Birds

What explanation can be given for the effects of pollution by organochlorine compounds on the reproductive functions of birds? An early theory was that the hormone balance is disturbed following stimulation of the hepatic microsomes. Even quite a few years ago it was possible to show that a number of toxic substances cause a proliferation of the endoplasmic reticulum in the hepatocytes of birds and mammals. Associated with this proliferation is a stimulation of the production of microsomal enzymes — hydrolases and oxydoreductases — responsible for hydrolysing and oxidizing toxic substances. In addition, these enzymes will also accelerate the catabolism of various organic substances. They will cause increased hydroxylation of steroids, particularly oestrogens and androgens. For instance, Peakall (1967, 1968) demonstrated that the hepatic microsomes of pigeons treated with DDT produced a larger quantity of polar metabolites (hydrosoluble) of oestradiol, progesterone and testosterone than that found in normal birds.

It is easy to understand how a perturbation of the hormone balance following increased catabolism of oestrogens can alter the fertility of birds and the mechanisms involved in eggshell secretion, both of which are controlled by these hormones. Also, certain organochlorine compounds such as DDE inhibit carbonic anhydrase, the enzyme which governs the depositing of calcium in the shell.

The phenomena of eggshell thinning observed with DDT, dieldrin and other chlorinated cyclodienes as well as certain PCBs must be attributed to cumulative effects from these different physiological disorders. However, it does not seem that increased catabolism of the various sexual hormones by organochlorine compounds is the only endocrinological cause of the problems observed. Until now, physiotoxicologists have almost systematically ignored the possibility of a malfunction of the cerebral cortex–hypothalamus–hypophysis axis, following chronic poisoning by organochlorine compounds, with resulting irregularity of all the endocrine functions of the contaminated organism.

Various experiments suggesting the existence of such effects have been published over recent years.

Gellert *et al.* (1972) showed that o,p'DDT and one of its principal metabolites, DDA (2, 2-bis parachlorophenyl acetic acid), administered to rats, exercised an oestrogen-like action by stimulating the uterine and vaginal mucous membranes and reducing the seric LH. These researchers considered that such an effect must be related to a negative feedback from these toxic substances to the hypothalamus or the hypophysis (pituitary).

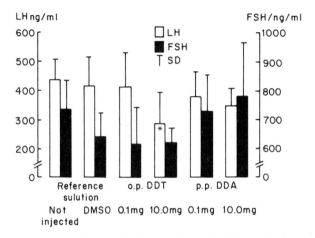

Fig. 3.14 Effects of contamination by DDT and certain of its catabolites on the secretion of pituitary gonadotrophic hormones in female rats that have had their ovaries removed. There is a significant reduction in LH, proving that DDT has an oestrogen-like action. SD = standard deviation. (After Gellert et al., 1972. ©by the Endocrine Society)

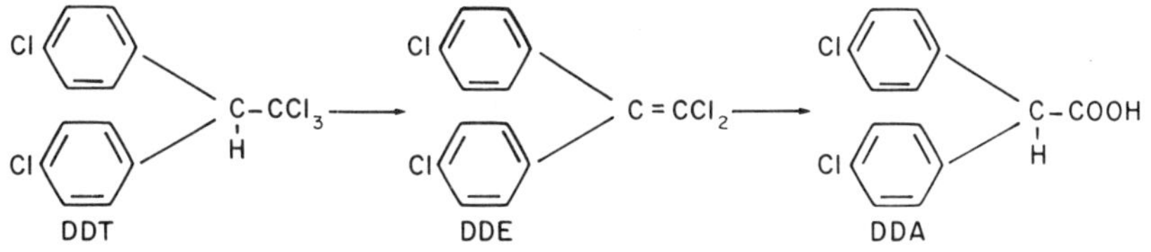

Fig. 3.15 Metabolic stages of DDT biodegradation in mammals. It is first converted to DDE and then to 2, 2-bis parachlorophenyl acetic acid (DDA), which is hydrosoluble

Our own research has shown modification of adrenal activity in mice exposed to low dosage of lindane (Ramade and Roffi, 1976) during sublethal uptake of this insecticide. There was a significant increase in the weight of the adrenal glands of the contaminated animals associated with a cortical hypertrophy of the glands. On the other hand, chronic uptake of dieldrin (for more than 2 months) causes a ponderal decrease in the adrenal glands with atrophy of the cortex. Comparable effects are caused by orally long-term exposure (56 weeks) of mice given food containing 16 and 65 ppm of lindane, even though these concentrations cause no mortalities among the dosed mice (Roffi and Ramade, 1981). In the exposed mice there is a hyperplasia of the glomerulus zone in the cortex of the suprarenal gland, accompanied by a significant increase in the amount of adrenaline in the medulla of those mice that have been given 65 ppm of lindane in their food. (cf. Fig. 3.16).

Similarly, Jefferies et al. (1971) noticed a hyperthyroidism and then a hypothyroidism in pigeons fed with increasing doses of DDT.

All of the different explanations of these observed modifications to endocrine activity imply a change in the processes of pituitary hormone production under the influence of the insecticide.

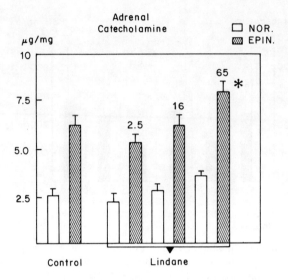

Fig. 3.16 The influence on the adrenaline catecholamines of increased amounts of lindane added to mice's food.
Nor = noradranaline; Epin = epinephrine
A significant difference is observed for epinephrine, in relation to the control substance, at a concentration of 65 ppm of lindane in the food.
(After Roffi and Ramade, 1981)

Later research involving contamination of pigeons by doses of 18, 36 and 72 mg/kg/day of DDE and 1, 2 and 4 mg/kg/day of dieldrin has all shown that these substances have a goitrogenic effect on the thyroid of the birds. Their glands are enlarged and contain less colloid, their follicles are reduced in size compared with the control birds and show an epithelial hyperplasia associated with vascular congestion (Jefferies *in* Moriarty, 1975).

A final cause of the reduction of animal populations due to organohalogen compounds is the increase in perinatal and juvenile mortality among offspring of females that have been exposed to these substances. This phenomenon is very evident in bird populations. When the vitelline sac is reabsorbed during the last days of embryonic life and the first few days after hatching, there is a corresponding circulation of the organochlorine compounds which may have accumulated particularly in the vitellus as it is rich in lipids.

In the same way, there is a high mortality rate among fish fry hatched from contaminated eggs just at the stage of resorption of the vitelline vesicle.

With species of insectivorous vertebrates, the young animals with a high metabolism and whose food is polluted by organochlorine residues are particularly vulnerable to insecticides. Examples abound in ornithological literature, including that of the high mortality rate in the nest of insectivorous Passeriformes chicks following the use of organochlorine insecticides in their territory.

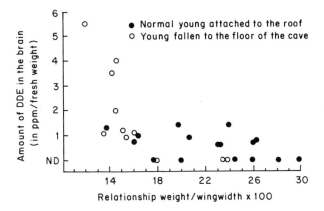

Fig. 3.17 The correlation between the general condition of young bats (expressed using the parameter of weight/wing width × 100) and the amount of DDE in their brain. This diagram suggests that the young bats which have fallen to the floor are thinner than those remaining in the normal position and also that the amount of DDE in their brain is higher. (After Clark *et al.*, 1975. *Reproduced by permission of the American Society of Mammalogists*)

It also appears that the dwindling numbers of bats in North America and Europe can be partly attributed to the fact that the young are fed on insects containing organochlorine residues. In the colony of *Tadarida brasiliensis* found in Bracken Cave, Texas, Clark *et al.* (1975) noticed that an unusual proportion of the young bats were incapable of remaining attached to the walls and roof of the cave. They were falling to the floor to be quickly devoured by the saprophagoous and necrophagous animals living there. A study of the correlation between the size of the young bats and the DDE residues found in their brain suggested that the ones which were falling to the floor were more contaminated. The cause of this higher mortality rate would be the pollution of the insects making up the bats' diet.

More recently, Clark (1981) concluded that organochlorine insecticides have caused an alarming drop in the populations of *Myotis grisescens* in Missouri and *Tadarida brasiliensis* in New Mexico. More generally, he considers that organohalogen compounds are a serious threat to certain Chiroptera populations of major importance and that these substances could also have been a determining factor in the past in the decline of other populations of North American bats.

IV—The Action of Organohalogen Compounds at the Cellular Level

The organohalogen substances being studied here are chlorinated or fluorinated hydrocarbons. They are all more or less strongly apolar molecules and there fore hydrophobic and lipophilic. This means they have a strong affinity for cellular membranes, which are rich in lipids. It is certainly not by chance that in almost all cases, organochlorine compounds behave like membrane poisons.

Their liposolubility also makes them attracted to nervous tissue whose cellular membranes contain many complex lipids, and this explains some of their neurotropic properties.

It has been possible to prove that DDT fixes itself on the neurone membranes and disrupts the nerve function by blocking axonal transmission. Lindane and chlorinated cyclodienes, which are other powerful neurotoxic substances, also act on nerve conduction by mechanisms which are as yet not fully understood.

Both DDT and most of the chlorinated cyclodienes are efficient inhibitors of the membrane $Mg^{2+}-Ca^{2+}$ and Na^+-K^+—At Pases, which explains why these substances interfere with axonal transmission.

It is probably also due to lindane's affinity for cellular membranes that proliferation of the lysosomal system is seen in various cells that have been polluted by this insecticide (cf. Ramade, 1966; Roux, 1973 in Boulekbache et al., 1974). Lindane also perturbs cell division. It blocks the polar ascension of the chromosomes and so bears a certain similarity to colchicine. Levain and Puiseux-Dao (in Boulekbache et al., 1974) showed that in this way it stopped the normal process of anaphase and telophase in a unicellular alga (Dunaliella euchlora). Separation of the daughter cells remains incomplete, resulting in the formation of giant plurinuclear cells. At a concentration of 5 ppm, all growth of this phytoplanktonic species is inhibited. Similarly, this insecticide disrupts mitosis in the root cells of Allium, with anomalies in the make-up of the spindle and incomplete formation of the phragmoplast. Such alterations to the karyokinesis result in chromatid breakages and, therefore, chromosomal mutations.

In the giant unicellular alga Acetabularia, inhibition by lindane of cell multiplication is accompanied by a slowing-down of morphogenesis. The ways in which this insecticide perturbs cell division are still very little understood. In fact, even though DNA synthesis seems to be affected in cultures of animal cells, in the cells of phanerogams the incorporation of [^{3}H]thymidine remains normal (Puiseux-Dao et al., 1977). It is possible that lindane behaves like a membrane poison and disrupts synthesis of the tubulins that play such an essential role during mitosis.

Numerous organohalogen compounds are capable of causing genetic mutations. Use of the Ames test (cf. Ch. 1, page 24) has provided proof of the mutagenic property of many of them some being important substances. this is true of vinyl chloride and vinylidene chloride used in the production of plastic materials and also of 2-chlorobutadiene, a basic monomer used in producing synthetic rubber.

V — Pathological Consequences Inherent to Environmental Pollution by Organohalogen Compounds

The mutagenic potential of various organohalogen compounds poses a worrying health problem. It is now known to be true that the vast majority of mutagenic compounds are also carcinogic and vice versa.

Vinyl chlorides and vinylidene chlorides are metabolized by the enzymes in the hepatic microsomes into epoxides which induce cancers. The Ames test has furnished proof *a posteriori* of their mutagenic and, therefore, carcinogenic power. In fact, even quite a few years ago it was noticed that there was a proliferation of cases of hepatic angiosarcomas—a malignant tumour in the liver which used to be very rare in human populations—among workers cleaning out PVC polymerization vats. More recently, chlorinated vinylidene has been shown to be carcinogenic in experiments with animals.

Other research has demonstrated that there is a certain carcinogenic potential in various organochlorine insecticides. Turosov *et al.* (1973) found, for example, that the exposure of six consecutive generations of mice to concentrations in their food of DDT between 2 and 250 ppm caused a significant increase in the incidence of hepatic tumours, even when the concentration was very low for the male mice and from 10 ppm for the females. A particularly malignant form of tumour, called a hepatoblastoma, was found to occur eight times more frequently in those mice exposed to 250 ppm than in the control animals.

Similarly, Fitzhugh (1964) showed that when aldrin and dieldrin are included at a concentration of 10 ppm in mice's food, after two years it causes hepatomas four times more frequently in the subject mice.

In fact, out of a total number of 25 organochlorine pesticides (insecticides, miticides and fumigants) produced in the United States during the last decade, no less than 19 of them were found to be carcinogens in animal experiments (according to Epstein, 1981).

It is very fortunate that, for the most part, organochlorine insecticides are weak carcinogens. Taking into account the serious lack of experimentation preceding its large-scale use, 'a substance like DDT could have been a carcinogen whose effects would have been paid for world-wide' (Fournier, 1974).

Various organochlorine compounds have teratogenic properties. In their *in vitro* experiments, David and Lutz-Ostertag (1972, 1973) noticed that DDT causes anomalies in the embryogenesis of the quail. These anomalies consist of malformations of the genitalia: the partial or total involution of the Müller duct in the females, partial feminization of the testicles and retention of the right female Müller duct in the males and a significant reduction in the number of gonocytes and in the thickness of the ovarian cortex (Lutz-Ostertag and Lutz, 1974).

VI—Global Effects of Pollution by Organohalogen Compounds

1. Action on primary production

Organohalogen compounds can affect primary production in both limnic and marine environments. They are in fact extremely toxic to phytoplankton. For instance, toxaphene—a chlorinated camphene related to insecticides of the cyclodiene group—at a dose level of 7 ppb stops all growth of a phytoplankton comprising a mixture of species of the genera *Dunaliella, Monochrysis,*

Protococcus and *Phaeodactylum*. When a planktonic community of *Platymonas* and *Dunaliella* spp. were exposed to 1 ppm of various substances for 4 hours, the productivity decreased by 29% with lindane, 77% with DDT and 95% with chlordane (*in* Ramade, 1968).

Wurster (1968) measured carbon fixation activity using the ^{14}C bicarbonate method in normal cultures and DDT-contaminated cultures of various species of marine diatoms and dinoflagellates (*Skeletonema, Pyramimonas, Peridinium* and *Coccolithus*). He discovered a decrease in photosynthesis activity once the concentration of the insecticide in sea-water reached 10 ppb. At 50 ppb, carbon fixation by the culture went down by 50%.

Fisher (1975) found that not only DDT but also PCBs (at 10 ppb) can significantly reduce the primary production of two species of marine diatoms. Interestingly, he shows that this reduction does not come from a perturbation of the cellular mechanisms in photosythesis but from a decrease in the mitotic rate of the species and a corresponding drop in the total photosynthetic activity of the culture, with activity of the individual cells remaining unchanged.

The primary production of *Skeletonema costatum* is particularly affected by DDT. A concentration of 50 ppb reduces by 92% the number of cells per millilitre of the culture after 48 hours, affecting, therefore, production of the plankton by the same proportion. For the diatom *Thalassiosira pseudonana*, 10 ppb of PCB lowers the number of its cells by 40% compared to the control culture.

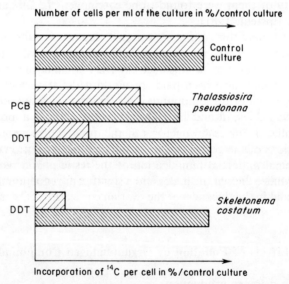

Fig. 3.18 The influence of DDT and PCBs on two species of marine diatoms. Photosynthetic activity remains constant at a concentration of 50 ppb but the number of cells per millilitre of the culture is considerably reduced after 48 hours of action by the toxicant, resulting in a decrease in net productivity compared to the control culture.
(After Fisher, 1975, *Science*, **189**, 463–464. *Copyright 1975 by the AAAS*)

As a general rule, concentrations of PCB between 1 and 10 ppb cause a noticeable decrease in the total biomass as well as cellular size when they are added to *in vitro* cultures of phytoplanktonic populations from marine littoral zones. The same PCB concentrations also inhibit chlorophyll synthesis from between 3 and 4 days without, however, perceptibly changing photosynthetic activity (O'Connors *et al.*, 1978). Fisher's experimental results (1975) have been confirmed by later investigations. For instance, Harding and Phillips (1978) showed the inhibiting effects of PCBs on the frequency of cell division and on incorporation of [^{14}C]carbonate by phytoplankton. However, it should be noted that the concentrations of PCB used, even though they can be found locally in coastal waters near industrial zones, are about 100 times higher than the average concentration of PCB found in marine surface waters.

Another worrying factor is the possibility of synergistic effects between PCBs and DDE present in the oceans. As an example, although 10 ppb of PCB or 100 ppb of DDE do not significantly affect the growth of a culture of the diatom *Thalassiosira pseudonana*, a mixture of these substances at the same concentrations causes such a slowing-down of cell division after 4 days that the density of the diatom in culture is 20 times less than in a control culture (Hunding and Lange, *in* Butler, 1978).

Extrapolation of these various experimental results to the real marine environment does of course lead the specialists to different conclusions. For some of them, the presence of insecticide and PCB residues in all marine organisms (cf. page 99) proves the existence of a level of contamination which could effectively lower primary production in the world's oceans. For others, this contamination will have no effect because strains of phytoplankton resistant to these low doses of organochlorine compounds have already been produced in the laboratory and have even been observed in coastal water and should, therefore, develop spontaneously in polluted waters (Mosser *et al.*, 1972).

Fisher (1975) does, however, rightly point out that a proliferation of phytoplanktonic strains resistant to organochlorine compounds would mean many biocoenotic upheavals. Even assuming that total primary productivity in the oceans were not affected, the decrease in species of phytoplankton sensitive to organochlorines would still alter the structure of consumer communities constituting the marine zooplankton and nekton — herbivores and carnivores — situated higher up the trophic systems. Also, development of resistant strains able to accumulate larger quantities of organochlorine compounds without any apparent physiological damage will result in a mounting concentration of these substances in food chains. The combined actions of all of these phenomena would be bound to have negative consequences on secondary productivity in the oceans.

2. Effects of pollution by organohalogen compounds on biogeochemical cycles

Some people have said that there is no foreseeable risk of reaching concentration levels in sea-water that might affect primary production. This argument is based on the fact that the saturation level of DDT in solution is 1 ppb in water, which is well below levels that disrupt production.

In reality, this theory does not stand up to scrutiny. The potential for reaching phytotoxic levels does exist. Concentrations of DDT and PCB above 100 ppb have already been found in sea-water. This comes from the fact that the upper layers of the ocean — where phytoplankton are concentrated — contain varied organic matter capable of emulsifying apolar compounds that are insoluble in water. Also, there is the possibility of synergism between pollutants which favours dispersion of organohalogen compounds.

Hydrocarbons dumped at sea (and unfortunately also on the continental shelf, even near coasts) by oil-tankers in the process of rinsing ships' holds, dissolve traces of organohalogen compounds suspended in the first few centimetres of the ocean and in this way enable them to emulsify in the surface layers.

There could eventually be an imbalance in the C/O_2 ratio if organohalogen pollution of the oceans causes a reduction in primary production and thereby a long-term disruption of the biogeochemical cycles of carbon and oxygen. However, in spite of the existence of some potentially dangerous factors, it must be admitted that such an eventuality is still largely hypothetical in the absence of any well-proven experimental data.

Nevertheless, pollution of the biosphere by organohalogen compounds does raise fundamental questions in addition to the immediate concerns. They are of a socio-economic, if not ethical, nature and are directly related to the structure of our industrial civilization. Firstly, no scientist could accept the continued commercialization far and wide on a massive scale and in vast quantities of substances whose ecotoxicological properties have not been properly assessed. We cannot be satisfied *ad infinitum* with the often partial withdrawal *a posteriori* of the most dangerous substances to public health. It is absolutely intolerable that a sort of technocrat–industrial collusion maintains a true conspiracy of silence about certain pollutants, and not insignificant ones. . . .

As a form of conclusion, it is worth reading the following comments by Risebrough (Laboratory of Marine Resources, University of Berkeley, California), one of the best specialists on pollution by organochlorine compounds:

> The DDT producers and their sympathizers have clearly shown a lack of public responsibility, . . . The Monsanto Company, by withholding information about production figures and PCB use, is also clearly acting against the public interest. The deficiencies uncovered in the Pesticides Regulation Division of the Department of Agriculture indicate that government agencies will not always put public interest before private interest. The continuing need is for solid data that will answer the pertinent questions and for impartial, independent interpretations of these data by scientists . . . whose conclusions are not influenced by political or commercial considerations. Scientists in a university or other private capacity are essential for the fulfillment of this responsibility and they, by default, must assume the major part of this responsibility.

After Risebrough *in* Hood *Impingement of man on the ocean.* Wiley Interscience, 1971, p. 281.

B—MERCURY

The toxicity of mercury has been known for a long time. Historically, this metal was even one of the first pollutants giving rise to a legal conflict in the industrial era: in 1700, the inhabitants of the town of Finale, in Italy, took out an action at law against a manufacturer of mercuric chloride whose emissions had poisoned many citizens.

More recently, the Minamata illness in Japan was a sad illustration of the very serious consequences to human health from the careless release into the environment of industrial effluents containing toxic substances, however diluted.

I—Principal Sources of Pollution by Mercury

There are several direct and indirect causes of contamination of the ecosphere by mercury.

1. Direct causes

The industrial uses of mercury often involve the partial or total discharge into the natural environment of wastes contaminated by this metal. Although mercury is easily recyclable from many of its applications, there are numerous other processes where its recuperation is either not considered or poses difficult technological problems.

For instance, the use of mercury as an electrode in the preparation of caustic soda generally results in the disposal of 250 g of the metal for every tonne of the alkali produced. As a little mercury gets into the soda, which is then used in certain food industries (cooking-oils, dairy products, as a neutralizer in tinned foods, or for chemically skinning fruits and tubers, etc.), it follows that the mercury can contaminate these foods.

When mercury is used as a catalyst in the chemical industry, some of it is then discharged in the liquid effluents. The main reason for the contamination of Minamata Bay was that part of the mercuric chloride being used as a catalyst in the synthesis of acetaldehyde by the Chisso chemical plant was released into the sea along with the waste waters from the plant (synthesis of 1 tonne of acetaldehyde releases some 30 to 100 g of mercury into continental or coastal waters).

Other industrial uses of mercury involve the complete dispersion of mercuric products into the environment. Examples include cosmetics, medicines with a mercury base (mercurochrome, mercryl®, etc.), fungicidal paints used to treat wood for the building industry (cladding, timber-frames, etc.).

Organomercury fungicides represent a significant cause of environmental pollution by mercury. Although the dangerous methyl mercury has been banned from use as a fungicide because of its extreme toxicity, various other organic mercury compounds are still being used to destroy mildew and different phytopathogenic agents. Aryl mercuries such as the phenylmercuric acetate and

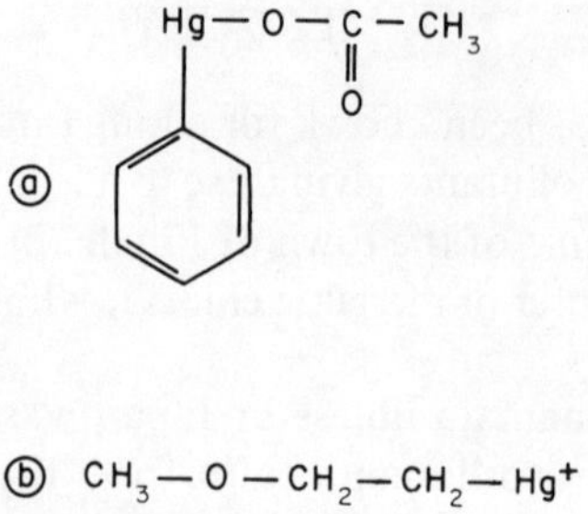

(a)

$$Hg-O-\underset{\underset{O}{\|}}{C}-CH_3$$ (attached to phenyl ring)

(b) $CH_3-O-CH_2-CH_2-Hg^+$

Fig. 3.19 Examples of organomercury fungicides:
(a) phenylmercuric acetate; (b) methoxyethyl mercury

alkyl mercuries (methoxyethyl mercury silicates or methoxymethyl mercury) are employed on a vast scale to protect seeds against various fungi which cause 'damping-off' of seedlings, and also to prevent mould forming on cellulose and paper pulps.

2. Indirect causes of contamination

One important cause of pollution of the biosphere by mercury is a result of the combustion of fossil carbon derivatives. The coal extracted in the Soviet Union contains an average of 0.28 ppm of mercury, and certain Californian crude oils contain between 1.9 and 21 ppm of mercury. Some researchers have estimated that up to 3000 tonnes of mercury per year are discharged into the atmosphere from using various fossil fuels. It is certainly not unreasonable to fix the total amount of mercury released into the biosphere by our technological civilization at 10 000 tonnes per year.

II.—The Biogeochemical Cycle of Mercury

In order to understand the mechanisms of mercury pollution it is necessary to know how it is circulated in the natural environment.

The biogeochemical cycle of mercury takes place between the three compartments of the ecosphere: the continents, atmosphere and lithosphere. Mercury is easily sublimed and goes into the air; also there are some very volatile forms of this element, such as dimethyl mercury, which aid its dispersion into the atmosphere. The quantities of mercury found spontaneously in the different ecosystems have two origins: volcanic eruptions and erosion by water which by leaching carries a fraction of the mercury contained in surface rocks into streams and rivers and so to the oceans. Jernolov (1969, 1972) showed that mineral mercury accumulated in sediments can be transformed by benthic bacteria into methyl mercury and then into dimethyl mercury with the reaction:

$$Hg \leftrightarrows Hg_2^+ \leftrightarrows CH_3Hg^+ \leftrightarrows (CH_3)_2Hg$$

This bacterial conversion of mineral mercury into organic mercury is a primordial factor in the field of ecotoxicology because aquatic and even

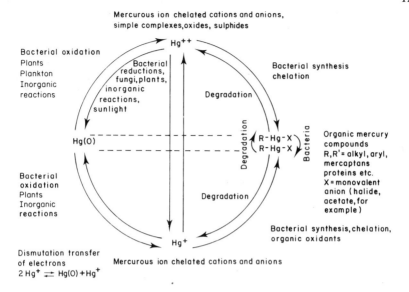

Fig. 3.20 Changes in the state of mercury in the biosphere as a function of the action of different biogeochemical factors. (After Goldwater and Stopford, 1977, *Environment and Man*, Vol. 6, Blackie, Glasgow and London.)

terrestrial food chains are mainly contaminated by alkaline derivatives of mercury, of microbiological origin, and principally by monomethyl mercury ($CH_3 Hg^+$). Fig. 3.20 is a diagram showing the intervention of different microbiological and chemical agents in the mercury cycle in nature. In practice, $R = CH_3$ in virtually all cases, even though the formation of monoethyl mercury ($C_2 H_5$) Hg^+ has been found in some Japanese rivers. Likewise, abiotic transmethylation, theoretically possible in the aquatic environment if stable alkylmetallic derivatives are present, seems to be an extremely rare phenomenon.

Microbiological methylation is a very slow process in the aquatic environment, with only 1% of the mercury contained in sediments undergoing this transformation per year. The most favourable environment for methylation is eutrophic sediments, rich in organic matter, at quite a high temperature (stimulating bacterial activity), with a positive redox potential enabling molecular oxygen to be diffused, and whose pH is slightly acidic. Consequently, lakes, hydroelectric reservoirs and some marshy and/or delta zones are the most favourable locations for the biotransformation of mineral mercury into alkyl mercury.

The very volatile dimethyl mercury leaves waters and goes into the atmosphere, thus explaining the contamination of lakes and other waters that have never been exposed to mercury-polluted effluents. On the other hand, monomethyl mercury ($CH_3 Hg^+$) stays in the hydrosphere where it is incorporated into food chains by the classic process of uptake by phytoplankton, on to zooplankton and finally into predators. When the plants and animals die, the mercury is returned to the sediments and the whole cycle starts again.

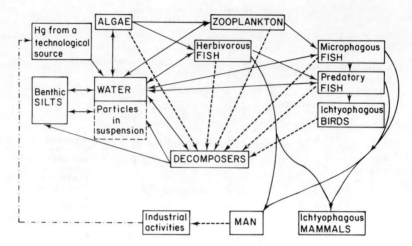

Fig. 3.21 Diagram of the circulation of mercury in aquatic ecosystems. (Adapted from Hartung, 1972.)

The speed with which mercury is incorporated into the biomass and its transfer into food chains varies quite significantly according to its chemical form. A comparative study of the transfer process was conducted *in vitro* between water, an alga (*Chlorella vulgaris*) and a cladoceran (*Daphnia magna*) using two compounds, mercuric chloride ($HgCl_2$) and methyl mercury chloride (CH_3HgCl). With all else being equal, the concentrations reached in the *Daphnia* were found to be 2.7 times higher with the CH_3HgCl than with the $HgCl_2$. In addition, three times more mercury was found to be incorporated into the food chain at a temperature of 18 °C than at 10 °C (Boudou and Ribeyre, 1981).

Mercury is normally present in minute traces in non-polluted atmospheres. Its concentration there will never exceed 0.0025 mg/m^3 of air, or less than 2 ppt (parts per trillion). However, it reaches much higher levels of concentration in industrialized zones. Brar *et al.* (1970), for example, found 0.04 mg/m^3 in the Chicago air, which is 20 times more than normal levels.

In continental and ocean water, there is an average concentration of mercury of the order of 1 ppb, and the level is 2 ppm in humus-rich soils.

Due to the very low biodegradability of certain organic derivatives of mercury (CH_3Hg^+ in particular), it tends to accumulate in living organisms. Algae can store it in their cells at levels 100 times higher than its concentration in water, and pelagic fish caught in remote areas of the oceans have been found to contain up to 120 ppb of mercury in their muscles.

III—The Contamination of Biocoenoses and its Consequences

The pollution of ecosystems by mercury from modern industry is well proven. Studies done over the last 15–20 years by Japanese, Scandinavian and American ecotoxicologists have shown irrefutably that when mercury is released into waters either in a mineral or organic form it will inevitably transform itself into methyl mercury, which is virtually non-biodegradable and accumulates in the organisms which make up different trophic systems.

1. Terrestrial ecosystems

These are mainly contaminated by mercury from the use of organomercury fungicides in agriculture. From 1960, various Swedish ecologists suspected that these substances were adding to the decline of the avifauna in their country.

The practice of using seeds that have been coated with organomercury fungicides seems to play a vital role in the contamination of agroecosystems. Otterlind and Lenerstedt (1966) measured 8 to 45 mg/kg of mercury in pigeons and 11 to 136 mg/kg in various finches caught in fields in southern Sweden. Similarly, of 70 goshawks and buzzards from various regions of Sweden, 67 of them contained abnormal amounts of mercury. Some tawny owls found dead in fields about 10 days after seed sowings had no less than 270 mg/kg of mercury in their liver and kidneys.

The work of Berg et al. (1966) demonstrated the determining role of organomercury fungicides in the decline of the Swedish avifauna. An analysis of the mercury content of feathers from numerous species of birds conserved from 1815 in some Swedish museums compared to recent specimens provided clear proof of their thesis. In fact, because the phanera of warm-blooded vertebrates (hair, feathers and other tegumentary parts) are natural channels

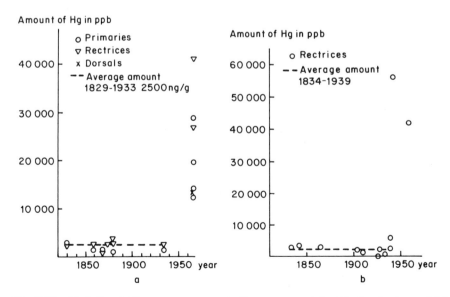

Fig. 3.22 Variation in the average amount of mercury contained in the feathers of (a) the eagle-owl and (b) the peregrine falcon in Sweden between 1830 and 1965. (After Berg et al., 1966. *Reproduced by permission of* Oikos)

for eliminating mercury from the body, the concentration there of the metal is a good indicator of the degree to which the whole organism is contaminated. This research showed that the amount of mercury found in the various specimens examined was remarkably constant over a period of more than 100 years before it suddenly went up from 1940 onwards, which was the year that alkyl mercury fungicides for treating seeds were introduced into Sweden.

This study also showed an excellent correlation between the mercury content of the feathers and the trophic level occupied by each species. There was an average content of 6 ppm in granivorous pheasants and partridges killed in 1960; it went up to 40 ppm in the eagle-owl, to 55 ppm in the peregrine falcon and to 60 ppm in the sea-eagle, all of which are strict carnivores. In contrast, the mercury content of the feathers of birds such as the willow grouse which live a long way from cultivated areas, did not present any significant variation since the beginning of this century.

Various more recent studies show that North American birds of prey are just as seriously contaminated by mercury and other heavy metals. Table 3.3 shows the amounts found by Snyder *et al.* (1973) in the eggs of the Cooper's hawk (*Accipiter cooperi*) from New Mexico and Arizona.

Man, situated at the end of the food chain in agricultural ecosystems, cannot escape from the effects of pollution of the trophic systems by organomercury compounds. The accidental (or deliberate) use of seeds coated with organomercurial fungicides can lead to dangerous contamination of human food. The eggs of laying hens from Sweden contained an average of 25 ppb of mercury against 7 ppb in eggs from western Europe (Tejning, 1967), until the levels returned to normal once organomercurial fungicides were banned from that country.

Moreover there have been serious cases of mercury poisoning causing several deaths among people who had eaten meat from animals raised on coated seeds, both in developed countries as well as in the Third World (cf. for example Curley, 1971 and Dinman, *in* Hartung and Dinman, 1972).

Table 3.3 The heavy metal content in the eggs of the Cooper's hawk from New Mexico and Arizona (in ppm). (After Snyder *et al.*, 1973, *Bioscience*, **23**, 300–305. Copyright©1973 by the American Institute of Biological Sciences)

Pollutants	Number of eggs		Average content in eggs containing detectable doses
	with detectable metals	with detectable metals	
Mercury	23	1	0.023 ± 0.003
Copper	23	0	0.63 ± 0.107
Lead	11	12	0.193 ± 0.044
Cadmium	20	4	0.12 ± 0.023

2. The contamination of limnic ecosystems by mercury

This is of equal concern as mercury can now be found in all organisms populating the continental waters of industrialized countries.

Observations

In Sweden at the beginning of the 1960s, the public health authorities had to ban fishing and the sale of fish caught in the lakes of more than 80 districts because they contained a higher level of mercury than that allowed in human food (0.5 ppm) by World Health Organization experts.

Similar considerations led the Canadian Food and Drug Administration and the United States FDA to prohibit the sale of fish from Lake St Clair, the Ottawa River downstream from the city of Ottawa and the St Lawrence River downstream from Cornwall. In addition, and for the same reason, commercial fisheries for various perches and eels were also closed in 1970 on Lake Erie and the southern part of Lake Huron.

In Lake St Clair, contaminated by the industrial effluents of Detroit, the average mercury content of fish can be between 0.3 and 5 ppm, most of it in the form of methyl mercury.

Mercury very quickly reaches higher concentrations than normal in contaminated limnic food chains. In Japan, although the waters of the River Agano, which were taking methyl mercury-polluted effluents from an acetaldehyde plant, contained no more than 0.1 ppb of mercury, there was a concentration of 10 ppm in the phytoplankton and 40 ppm in the fish, which is a concentration factor of 400 000.

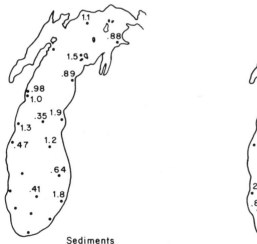
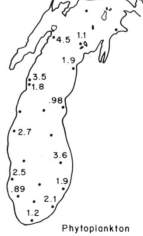

Fig. 3.23 Mercury pollution of the Great North American Lakes. They are at present extensively contaminated by mercury. These maps show amounts of mercury (expressed in ppm) in sediments and phytoplankton at various points on Lake Michigan. (After Copeland, *in* Hartung and Dinman, 1972)

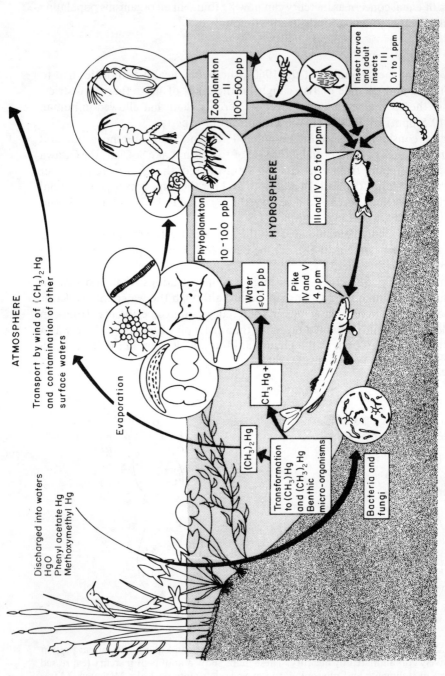

Fig. 3.24 Contamination of the food chain of the pike in Sweden (this diagram has been based on one by Duvigneaud (1974) and on analytical data on mercury contamination from various Swedish researchers). (*In* Ramade, 1982. *Reproduced by permission of McGraw-Hill*)

In Sweden, Johnels et al. (1967) found the same bioconcentration phenomena of methyl mercury in the food chain of pike (Fig. 3.24). An analysis of several dozen pike showed that they contained between 0.5 and 10 ppm of mercury and that concentration levels were higher in older fish.

Consequences

Pollution of freshwater fish by mercury causes a serious contamination of their predators, ichthyophagous birds, and, of course, fish-eating humans.

Table 3.4 lists the concentrations of mercury found in the muscles and liver of various ichthyophagous birds from Lake St Clair in the United States.

Similarly, fish-eating birds and other super-predators living on or near Lake Michigan show clear signs of mercury poisoning. Heinz et al. (1980) found an abnormal amount of mercury in water snakes (*Nerodia sipedon*) and garter snakes (*Thamnophis sirtalis*) sampled on Pilot and Spider Islands.

Research into the mercury content of the serum and urine of people living near the Great Lakes showed that those with a mainly fish diet had significantly higher levels of mercury than people who no longer ate fish (Wilcox, 1972). More than 33% of the residents in the Great Lakes region have unusually high mercury levels in their urine (Mastromatteo and Sutherland, 1972).

European limnic ecosystems also show worrying signs of mercury pollution. An analysis of principal organisms making up the trophic system of the River Rhine in Alsace showed the presence of mercury residues at each level (Kempf and Sitler, 1977). In addition, the ground waters of the Rhine have proved to be highly contaminated by mercury that has seeped in from the river-bed in the area upstream from Strasbourg.

All of the pike sampled in Alsace have no less than 0.29 ppm of mercury, with a maximum of 4 ppm found in a specimen from Munschhausen (*in*

Table 3.4 Contamination by mercury of ichthyophagous birds from Lake St Clair, United States. (After Dustman et al., 1972)

Species	Sex	Mercury residues (in ppm/fresh weight)	
		in the Muscles	in the Liver
Heron	♂	23.0	136!
(*Ardea herodias*)	♂	21.2	175!
	♀	8.3	66!
Common tern			
(*Sterna hirundo*)	♂	7.5	39!
Black tern			
(*Chlidonias niger*)	♂	0.61	3.5
Tufted duck	♂	1.2	3.4
Aythya affinis	♀	0.91	5.6

Carbiener, 1978). The fish-eating birds have an even higher level of contamination: 11 ppm in the kidneys and 18 ppm in the liver of the common tern (*Sterna hirundo*).

3. Contamination of marine biocoenoses

Marine animals caught near the coasts of industrialized countries and sometimes even in pelagic zones of the oceans seem for the most part nowadays to contain abnormally high concentrations of mercury.

Invertebrates

Molluscs are well known for their ability to accumulate various metals, even under natural conditions — the hepatopancreas of the scallop or *Pecten*, for example, can contain, 1°/00 of cadmium as a proportion of its dry weight. They are, therefore, able to act as bioconcentrators of the mercury present in their environment. Mussels gathered in the Thames estuary contained up to 0.65 ppm of mercury, and oysters from the same location contained 0.2 ppm. In Minamata Bay the mercury levels in various molluscs sampled amounted to between 11 and 40 ppm.

Vertebrates

Fish from the maritime zones of so-called developed countries also indicate considerable mercury pollution. For instance, cod from the North Sea contain an average of 0.175 ppm, and those caught in the southern Baltic have more than 1.3 ppm, whereas cod from non-polluted regions (such as the east coast of Greenland) contain scarcely 0.019 ppm of mercury (*in* Holden, 1973).

The French Mediterranean coast is a dumping ground for mercury from industrial waste. It is, therefore, not surprising that fish and molluscs caught in the Bay of Angels, into which the effluents from the city of Nice are poured, should be very contaminated. For example, the levels are 2.58 ppm of total mercury and 0.06 ppm of methyl mercury in the mussels and 2 ppm of total mercury in the chinchard fish (*Trachurus trachurus*) (of which 0.9 ppm is methyl mercury), etc. (After Aubert *et al.*, 1973).

Marine mammals living near the coasts of industrialized countries also have high concentrations of mercury. The grey seals (*Halichoerus grypus*) in Scotland contain an average of 10 to 50 ppm of the metal in their liver, and in some old seals as much as 720 ppm has been measured. Even more astonishing is the concentration level of 8.87 ppm found in the liver of a cetacean, the white dolphin (*Delphinapterus leucas*) caught near the coast of Hudson Bay, in a region far removed from any industry.

The Minamata Affair

The Minamata disease, which between 1953 and 1975 caused about a hundred deaths and disabled 798 other people (and even more in reality if certain unofficial information sources are to be believed), is a clear example of the possible consequences of modern technology's pollution of the oceans. The disease started in 1953 in Minamata Bay and its symptoms were various nervous disorders (sensory, motor and psychic). It particularly affected fishermen, who were of course big fish-eaters, and their domestic animals. There were cases reported of cats having hallucinations and then committing suicide by jumping into the water, which was certainly unusual given their well-known aversion to water.

It was eventually established that the sole cause of this 'epidemic' was the mercury being discharged (in the form of methyl mercury, $CH_3 Hg^+$) by the Chisso chemicals plant manufacturing acetaldehyde. The methyl mercury became concentrated in the marine trophic systems and reached high levels in molluscs, crustaceans and fish which were then eaten by local fishermen. Although the amount of methyl mercury in the sea-water (expressed as mercury) was no more than 0.1 ppb, it reached concentrations of up to 50 ppm in the fish, which is a concentration coefficient of 5×10^5. In some of the fishermen suffering from the Minamata disease, up to 528 ppm of mercury was found in their hair.

A little more recently (from 1962 onwards and especially during the 1970s) a population of Indians living in northern Ontario and with a diet mainly of freshwater fish, also contracted this disease, although their symptoms were less drastic than those of the Minamata disease. This case of mercury poisoning was caused by pollution of the rivers and lakes on the Indians' territory from mercury-contaminated waste released by papermills situated upstream.

IV — Physiotoxicological Effects of Mercury

Recently, the problem of the dangerous effects from low doses of mercury being continually ingested with food has given rise to some general physio-toxicological research, making considerable progress in understanding the reaction mechanisms and toxic properties of mercury.

Mutagenic Effects

On the cellular level, mercury has been proved to be mutagenic in both *in vitro* and *in vivo* studies. Ramel (1972) showed that alkyl and aryl mercury disrupt mitoses in the roots of garlic, and block them at the metaphase, causing the formation of C-mitoses. They can do this with far weaker doses than are needed of colchicine to produce the same effects.

In another study, Skerfving (*in* Goldwater, 1971) observed chromosomal mutations in the lymphocytes of fish-eating Swedes with a high level of mercury in their serum. The mutations were characterized by chromatid breakages, the appearance of extra-fragments, and even by chromosomes that had no centromere.

It is difficult not to relate these various mutagenic effects to the teratogenic effects observed in offspring of Japanese women affected by the Minamata

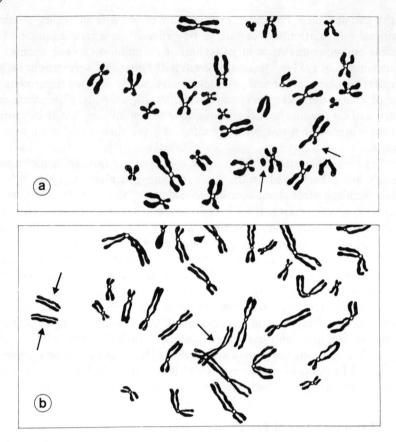

Fig. 3.25 Chromosomal mutations observed in the lymphocytes of fish-eating Swedes with a high level of mercury in their serum. The arrows on figure (a) show many chromosomal breakages with the appearance of extra fragments. The arrows on figure (b) show the abnormal chromosomes with no centromere. (After Skerfving *in* Goldwater, Scient. Amer., May 1971. p. 27)

disease, even if in most cases any abnormalities could be explained by the direct action of methyl mercury on the foetal brain.

Mercury Poisoning

Mercury poisoning mainly attacks the brain and kidneys, although mercury can also be accumulated in the liver.

Long-term exposure to methyl mercury results in serious disorders of the central nervous system, characterized by hearing difficulties, restriction of the visual field, *numbness in the extremities*; by motor disorders; trembling, *hesitating steps*, exaggerated reflex responses; and in some cases by psychic disorders. A histopathological study reveals a cortical atrophy in the cerebrum and cerebellum and an extensive pycnosis of the perikaryon of the neurones, which is very severe in the Purkinje cells, explaining the difficulties with balance experienced by sufferers.

It has also been possible to show that organic mercury inhibits glutathione-reductase and, to a lesser extent, the seric phosphoglucose isomerase serum.

Effects on Reproductive Functions

It has been proved that methyl mercury, in sublethal doses, can affect the reproductive functions of birds. Heinz (1974) noticed that when mallard ducks were fed for a year with 3 ppm of methyl mercury in their food, there was a decrease in the number of eggs laid, and an increase in embryo mortality and in deaths of the young chicks not long after hatching.

C—CADMIUM

Pollution by cadmium has become a serious environmental problem, even if up until now it is more localized and, therefore, less of a problem than mercury pollution.

Apart from having quite high intrinsic toxicity, cadmium also possesses significant carcinogenic potential.

Sources of Cadmium Pollution

Cadmium occurs naturally in trace amounts (from 1 to 250 ppb) in the surface rocks of the earth's crust, with an average of about 50 ppb. It is a by-product of the metallurgy of zinc and, to a lesser extent, of lead. The zinc ores from which it is principally extracted contain between 100 and 500 ppm before concentration. The current annual consumption of cadmium world-wide is slightly higher than for mercury (17 000 tonnes in 1975, of which 1460 tonnes was in France).

The main uses for cadmium are in electricity (storage batteries), in electronics (photovoltaic cells) and in metallurgy (surface-coating by cadmium electroplating). The plastics industry, with due allowance being made, also employs quite large quantities of cadmium stearate as a stabilizer in certain polymers. In West Germany, for instance, some 950 tonnes per annum are produced for this use.

There are, therefore, a number of causes of environmental pollution by cadmium at its metallurgical origin as well as at the stage of its industrial usage and its disposal after this use. For example, extraction of cadmium from zinc ore and burning of plastics stabilized by cadmium stearate are important sources of air pollution by cadmium. On the other hand, electroplating is a major cause of the cadmium contamination of continental waters when it is discharged into the local hydrological system in contaminated effluents.

There are also indirect causes of environmental pollution by cadmium. Like mercury, cadmium can be released into the atmosphere from the burning of fossil carbon derivatives. For example, lubricating oils and fuels contain from

0.07 to 0.54 ppm according to their nature and origin, and coal contains 1 to 2 ppm.

Chemical fertilizers of the superphosphate group can contaminate the soil as they have 0.05 to 170 ppm of cadmium according to their degree of purification.

The Transfer of Cadmium into Food Chains

Continental ecosystems can be contaminated by cadmium in two distinct ways: fallout of cadmium dust transported by the wind, and leaching into the soils from the use of impure phosphate fertilizers.

The cadmium enters plants through the roots or even by being absorbed through the leaves, not just via the stomata but also directly through the cuticle. The usual amount of cadmium found in the leaves of cultivated plants is from 0.1 to 2 ppm (in relation to dry weight).

Cadmium can also accumulate quite spectacularly in certain organisms in terrestrial trophic systems, and these organisms can then serve as bioindicators of the environmental pollution by cadmium. For instance, a terrestrial isopod, the woodlouse (*Oniscus asellus*) can concentrate cadmium in various organs, especially in soft tissues (muscles and fatty tissues) and in the hepatopancreas at levels that are considerably higher than those of the cadmium contained in the soil and litter of the forests where these animals live (Coughtrey et al., 1977). The concentration factor between the woodlouse's hepatopancreas and the litter is over 50.

Similarly, Sawicka-Kapusta (1979) suggests that deer antlers be used to evaluate the level of heavy-metal pollution (particularly by cadmium) of forest ecosystems. He compared the concentration of cadmium, chromium and lead, etc., in the antlers of the roe-deer (*Capreolus capreolus*) from various Polish forests. He was able to show that the amounts of cadmium and other heavy metals found in the deer sampled in Silesia, a region heavily polluted by industry, were far higher than those found in the antlers of roe-deer living in national parks in remote areas. There is, therefore, a clear correlation between environmental pollution by heavy metals and their concentration in deer antlers.

In general, the present levels of cadmium in human food are not worrying as yet, with the main danger coming from contamination of drinking-water. However, there are signs that the levels of cadmium polluting water and food are going up in industrialized countries. At the end of the 1970s, the people of West Germany (FRG) were ingesting an average of 0.476 mg per week of cadmium in their food, which is very close to the maximum limit allowed in human food (0.5 mg/person/week) established by the joint FAO–WHO committee (from Gerlach, 1981).

It was due to contamination by cadmium of drinking water that a few dozen people in Japan caught the disease called itaï itaï. Symptoms of the disease were bone abnormalities and necroses of various organs. The contaminated drinking

Table 3.5 The concentration of cadmium in marine invertebrates. (After Mullin and Riley, 1956.)

Phylum	Concentration (in ppm)
Molluscs	0.83 to 38
Echinoderms	0.24 to 15
Crustaceans	0.15 to 1.3
Cnidaria	1.2 (average)
Sponges	1.9 (average)
Protozoa	1.2 (average)

water contained 0.18 ppm of cadmium whereas non-polluted natural waters contain only 0.5 ppb.

Higher concentrations of cadmium are found in aquatic food chains. Even in areas unpolluted by industry, certain marine invertebrates, particularly molluscs, can accumulate significant amounts of cadmium in their organism. Mullin and Riley (1956) found a range of concentrations in the main phyla of marine animals, as depicted in Table 3.5.

Oysters sampled in remote parts of the New Zealand coastline contain 9 ppm of cadmium/fresh weight. The record concentration level in the animal kingdom is held by scallops (*Pecten*), with some 2000 ppm (compared with dry weight) found in its hepatopancreas (Bryan, 1976). Other analyses of the whole organism found 200 to 500 ppm of cadmium in the scallops/fresh weight.

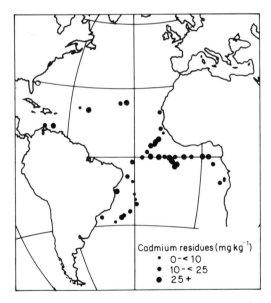

Fig. 3.26 The concentration of cadmium in the pelagic heteropteran *Halobates micans*. It is noticeable that a zone of maximum concentration corresponds to movements of the waters of the Atlantic. (After Bull *et al.*, 1977.)

One exceptional case of the bioaccumulation of cadmium by a marine invertebrate under natural conditions is that of the only true oceanic insect, the sea-skater (*Halobates micans*). This heteropteran, related to the Gerridae (pond-skaters), glides on the ocean's surface with its body completely out of the water except for tarsal paddles and is found in the inter-tropical waters of the Atlantic. These insects contain an average of 33 mg/kg of cadmium compared to their dried weight and even as much as 3000 mg/kg (Schultz-Baldes and Cheng, 1980). As these creatures live in areas far removed from any sources of pollution, their abnormal concentration of cadmium can be explained only by their diet, which consists of planktonic organisms floating on the surface, and also by the circulation system of the waters in the upwelling zones where they live. In fact, the high level of cadmium is due to deep waters, rich in mineral elements from the benthic silts, rising to the surface.

Limnic ecosystems are currently among the most exposed to cadmium pollution in industrialized countries. Studies on the Rhône basin (Teulon, 1973 *in* Comerzan, 1976) showed significant pollution of the trophic systems. Many samples of species of freshwater Teleostei analysed (bream, chub, roach and beaked carp) were quite highly contaminated, both in the Rhône itself as well as in parts of its unpolluted tributaries (Table 3.6).

Table 3.6 Cadmium pollution of cyprinoids in the Rhône basin. (After Teulon *in* Comerzan, 1976. Reproduced by permission of the Ministère de l'Equipement, du Logement, de l'Aménagement du Territoire et des Transports.)

Species	Concentration according to location of sample (maximum found in ppm)		
	Rhône (polluted)	Unpolluted tributaries Drôme	Ardèche
Bream	0.45	n.d.	0.30
Chub	0.25	0.1	0.05
Roach	0.30	0.1	n.d.
Beaked carp	0.70	n.d.	0.30

Physiotoxicological Effects of Cadmium

Cadmium has acute and long-term toxicity to mammals. It is particularly effective because it is not eliminated by the organism and so accumulates, mostly in the bones. There is also evidence of a correlation between cadmium poisoning and arterial hypertension.

This metal is also a powerful mutagenic agent. Numerous experimental studies and epidemiological surveys support this fact. For instance, Ramel and Friberg (*in* Friberg *et al.*, 1974) showed that adding 62 mg/kg of cadmium to the larval growing medium of drosophila caused the production of XO males, having lost their Y chromosome, in 0.3% of the specimens treated. Treating human leucocytes (white blood cells) in culture with cadmium sulphide causes, after 8 hours exposure, 14% chromosomal breakages, 3% formation of dicentric chromosomes and 6% translocation of extra fragments (Shiraishi *et al.*, 1972). Studies made in Japan of the chromosomal matching of patients suffering from itaï itaï disease showed that more than 30% of their cells had chromosomal mutations, whereas such abnormalities occur in less than 0.5% of the cells of normal individuals (*in* Friberg *et al.*, 1974; cited by Comerzan, 1976).

The role of cadmium as a primary carcinogenic agent is now well established in animal experimentation. Injected into different parts of the body, both cadmium oxide and cadmium chloride cause sarcomas in laboratory rodents. Its influence as a carcinogenic factor in humans is not yet proved, although according to some analyses, the cadmium content of tumour tissue is higher than in corresponding healthy tissue.

II — POLLUTANTS OF CONTINENTAL ECOSYSTEMS

These include all toxic agents whose effects are at present limited to terrestrial and limnic ecosystems, and where occasional contamination of the ocean by some of the substances is not yet sufficient to cause discernible ecotoxicological consequences.

A — POLLUTANTS OF AGROECOSYSTEMS

The use of various chemical products in agriculture can have different polluting effects not only on the agricultural ecosystems themselves, but often reaching far beyond them as biogeochemical factors cause their transport and dispersion to many terrestrial and limnic ecosystems.

I — Pollution by Pesticides

We have already looked at various aspects of pollution by pesticides, especially those whose ecotoxicological effects are of most concern at present, namely the organochlorines and the mercurial fungicides.

There are, however, many other pesticides in existence (more than 300 pesticidal substances are officially recognized for pest control use in France alone). In the United States, at the beginning of this decade, more than 1400 pesticides were counted (pure active matter), used in some 40 000 different commercial preparations (according to Epstein, 1981).

Some of the most widely-used **insecticides** at present are various organophosphorus compounds and *N*-methylcarbamates. The use of **herbicides** has grown steadily. Even though none of these substances is as stable as organochlorine insecticides, their dispersion into the countryside and sometimes over forests and limnic ecosystems (against vectors of parasitic and viral diseases) has significant ecotoxicological effects. These effects fall into two main groups:
— 'demoecological' effects on populations;
— biocoenotic effects on communities.

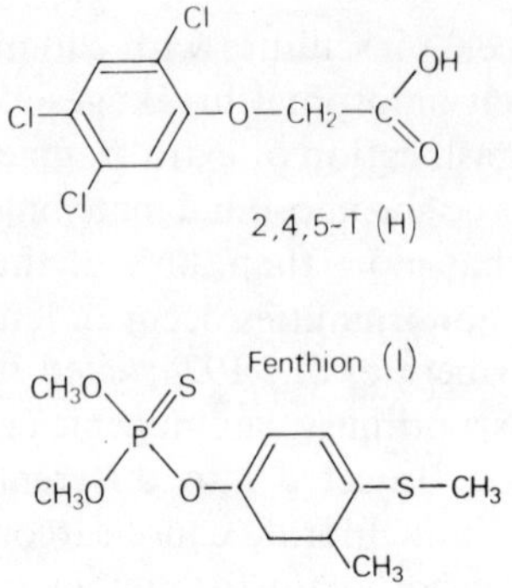

[Chemical structures: 2,4,5-T (H) and Fenthion (I)]

Fig. 3.27 Chemical formulae of two pesticides mentioned in the text. Fenthion is an organophosphorus insecticide used to kill the larvae of aquatic insects carrying parasitic diseases. 2, 4, 5-T is a defoliant from the group of phenoxyacetic acid derivatives; 2, 4-D, its dichloride homologue, is also used as a herbicide in cereal culture

1. Demoecological effects

These effects cause a decrease in animal or plant populations exposed to a pesticide and are the result of either immediate or long-term toxic action. They affect the populations by raising the mortality rate or lowering the birth rate of susceptible species and can even affect both rates simultaneously.

Organophosphorus Insecticides

Among the many different types of antiparasitic chemicals in use, organophosphorus insecticides are the principal cause of death from acute poisoning of the wild fauna. For instance, the use of phosphamidon to control the leafroller moth caterpillars on conifers (*Choristoneura fumiferana*) in Montana, USA, caused an 87% drop in the density of the bird population of a forest of *Pseudotsuga menziezii*. Out of 27 species counted before spraying, several had completely disappeared afterwards, particularly the tree creepers. A surviving breeding pair of the blue grouse (*Dendragapus obscurus*) was found after 15 days to have a level of seric cholinesterases 50% less than normal (Finley, 1965).

The replacing of organochlorine insecticides with fenthion and other organophosphorus insecticides in the fight against mosquitoes, which has taken place progressively since the beginning of the 1960s, is a considerable ecotoxicological advance. These newer compounds have far less persistence in water and a half-life that can be just 48 hours in some cases. However, there are certain limnic species, besides the target larvae and adult mosquitoes, that are very sensitive to these substances. For instance, fenthion in granular form at a dose level of 0.025 ppm is considered to be non-toxic to Copepoda, Ostracoda and Oligochaeta (*Tubifex* sp.) and also aquatic gasteropods. Nevertheless, Cladocera (*Daphnia* sp.) are quickly killed by such a dose as their lethal threshold is as low as 0.000 65 ppm (Muirhead-Thomson, 1971).

Although not as hypersensitive as the *Daphnia* to fenthion ($LC_{50} = 0.09$ ppb according to Ruber, 1965—Muirhead Thomson, 1971), Gammarides are also highly vulnerable to this pesticide (LC_{50} after 96 hours equal to 9 ppb).

Herbicides

It was the Vietnam war and large-scale usage of defoliants during the conflict that prompted serious studies of the ecotoxicological consequences of herbicides.

(a) Phenoxyacetic acid derivatives—Certain trees are extraordinarily sensitive to phenoxyacetic acid derivatives (2, 4-D and 2, 4, 5-T, for example). The dominant species in mangrove swamps (*Rhizophora* sp., *Avicenia* sp., *Brugueria* sp.), trees in tropical mountain forests (*Pterocarpus*, *Lagerstromia*), and arborescent Caesalpiniaceae were virtually eliminated from vast areas by defoliation programmes during the Vietnam war (*in* Westing, 1984).

The synthetic herbicides used in agricultural ecosystems are also very efficient. For instance, the systematic treatment of the regions of cereal culture in Europe have left them almost denuded of any wild plants, as is shown by the bare fields after harvest.

It also appears that phenoxyacetic acid derivatives, controversially used as defoliants in forestry, may in fact have some effects on various animal species, even in low doses. They also have a longer persistence time than is generally thought by their users. A population of birches in Sweden treated with 2, 4, 5-T still had residues of more than 10 ppm of the compound on their leaves two years after they were treated. (Erne, 1974).

Herbicides of this group can play an important part in the life of soils by modifying indirectly the invertebrate population through their effects on vegetation. Their toxicity to the edaphic zoocoenosis is not yet well known, although it seems to be weak. It has been shown, however, that 2, 4, 5-T affects the activity and longevity of the collembolan *Onychiurus quadriocellatus*.

Phenoxyacetic acid derivatives may only rarely be acutely toxic to homoeothermic vertebrates, but this is not true for poikilothermic animals, particularly limnic species. Toxic effects on freshwater fauna have been found at concentrations of between 1 and 4 ppm. Lhoste (1959) noticed a high mortality rate among aquatic insects, molluscs and various crustaceans exposed to a mixture of 2, 4-D and 2, 4, 5-T diluted in water to levels of 0.1 to 3.3 ppm. Research done in our own laboratory in Orsay (Seugé *et al.*, 1978), was able to establish a number of LC_{50}s after 96 hours (at 20 °C) of a mixture of 2, 4-D and 2, 4, 5-T (in a proportion of two-thirds to one-third). The LC_{50} for larvae of the *Chaoborus* sp. was 4 ppm; for the gasteropod *Lymnea palustris* it was

1.1 ppm; it was 0.75 ppm for mayfly larvae (*Cloeon* sp.); and it was 0.5 ppm for the amphipod *Gammarus pulex*.

Phenoxyacetic acid derivatives are also relatively toxic to fish. The LC_{50} after 24 hours is 0.5 ppm with 2, 4-D for freshwater Teleostei, and about 1 ppm for a mixture of 2, 4, 5-T and 2, 4-D.

In general, it would seem that pollution of water from the use of these defoliants in agriculture and forestry is fairly insignificant. However, Erne (1974) found traces of these substances in 25% of 600 samples taken from streams running through forests and cultivated zones in Sweden. In 10% of the samples, the concentrations were equal to or above 0.01 ppm, which is the minimum limit that can cause acute or subacute toxic affects in certain sensitive species (Cladocera, for example); in addition, a dozen of the samples contained more than 1 ppm (Fig. 3.28).

(b) Diuron—Other herbicides can be toxic to limnic fauna. Even urea substitutes like diuron (LC_{50} after 48 hours between 3 and 20 ppm for fish) can

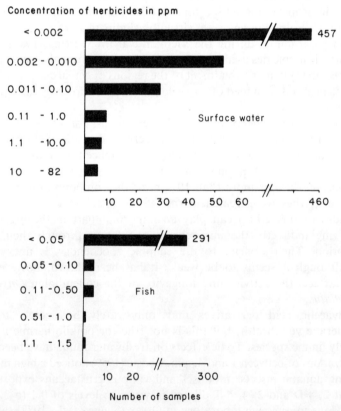

Fig. 3.28 The scale of contamination of Swedish limnic ecosystems by 2, 4, 5-T used as a defoliant in forestry. Although the majority of water samples bear only a trace of these substances, about 5% of those analysed contain levels which can have significant ecotoxicological effects. (After Erne, 1974)

cause acute or long-term poisoning. There was 100% mortality among an experimental batch of carp exposed for 40 days to a dose of 0.5 ppm. Also, if the same species of fish is raised in water constantly containing 0.1 ppm of diuron, they develop granulomatous lesions of the endocardium (Schultz, 1973).

Diuron can accumulate in limnic trophic chains. Koeman *et al.* (1969) found an average of 2.9 ppm in aquatic invertebrates raised in a pond contaminated with 0.4 ppm of diuron, and an average of 34 ppm in the tissues of the carp, which is a concentration factor of 85 times.

(c) Paraquat—Some research done *in vitro* (Streit, 1979) showed bioaccumulation of this herbicide in limnic invertebrates. The mollusc *Ancylus fluviatilis* and the leech *Glossosiphonia complanata*, grown in water containing 0.5 ppb of paraquat, accumulated 10 ppb in their soft tissues.

(d) Teratogenic properties—Phenoxyacetic acid derivative herbicides also display teratogenic properties. They were demonstrated particularly during the Vietnam conflict: there were congenital malformations in the offspring of pregnant women living in zones that were heavily treated with defoliants.

Various research has in fact proved that most of these abnormalities could be attributed to the presence of an impurity, dioxin (cf. later), in the defoliant used called 'agent orange'. Analyses showed an average level of 1.9 ppm of dioxin, with a maximum of 47 ppm in the 2, 4, 5-T used in Vietnam. There is experimental proof of the teratogenic qualities of the herbicides of this group. For instance, Båge *et al.* (1973), using a 2, 4, 5-T containing less than 1 ppm of dioxin, showed that these substances cause embryotoxic and teratogenic effects in mice. Having administered 110 mg/kg by intraperitoneal injection to pregnant mice, these researchers found there was a significant increase in the number of spontaneous abortions, an abnormal frequency in skeletal malformations (25% of the offspring were affected), and the occurence of renal and subcutaneous haemorrhaging, etc. 2, 4, 5-T administered on its own appeared to be more toxic than a mixture in equal parts with 2, 4-D. Lutz and Lutz (1970, 1973) and Didier (1974, 1975) showed that reputedly pure 2, 4-D and 2, 4, 5-T could have a teratogenic effect on bird embryos tested *in vitro*. The abnormalities observed in these studies, particularly in the organogenesis of the genital tract, caused by the herbicides in question must have demoecological consequences as they are obviously affecting the biotic potential of the contaminated animal populations.

It should be noted, however, that such effects are only caused in hen and quail embryos with quite high concentrations (intra-allantois injection of 2, 4, 5-T in concentrations of about 100 mg/kg, Didier, 1981). In the same way 2, 4, 5-T is teratogenic to rodents at concentrations of about 35 mg/kg (*in* Hay, 1980).

2, 4-D also acts on postembryonic development, and if it is absorbed over a prolonged period in concentrations of 10 ppm or more in food, it has a clearly inhibiting effect on the growth of young birds (Whitehead, 1973).

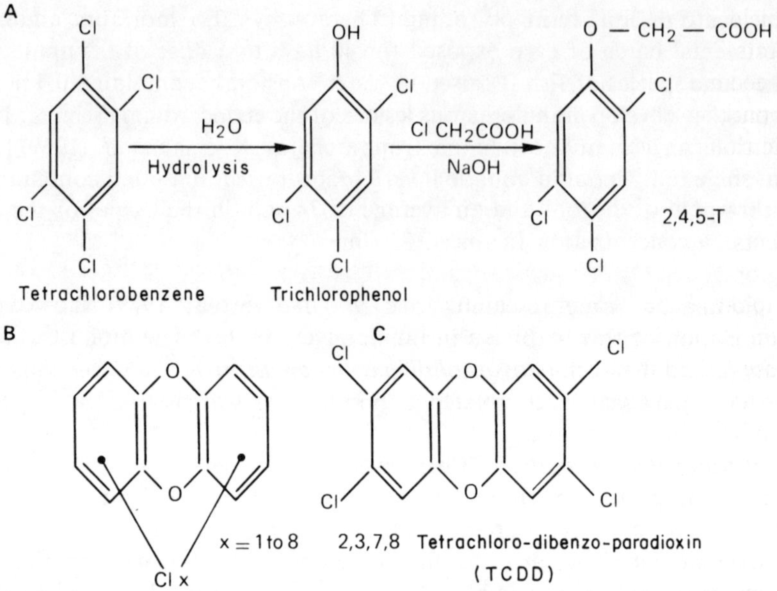

Fig. 3.29 A: Principal stages in the synthesis of 2, 4, 5-T. The dioxins, whose general formula is shown in B, are formed during the hydrolysis reaction of tetrachlorobenzene to trichlorophenol. C shows the most dangerous dioxin, TCDD

Dioxin

Various investigations into the toxicity of 2, 4, 5-T have shown that many of its toxic effects could be attributed to an impurity, dioxin, which is formed during the synthesis of trichlorophenol (cf. Fig. 3.29A and B). Trichlorophenol is a widely-used base substance in the pesticides industry because, besides its use in 2, 4, 5-T, it is used in the manufacture of two insecticides (trichloronate and *Ronnel*) and a herbicide (not commercialized in France), *Herbon*. Trichlorophenol is also employed in the synthesis of hexachlorophene.

Several dioxins can appear during the reaction leading to trichlorophenol (and also in that of dichlorophenol). The most important dioxin ecotoxicologically is TCDD (2, 3, 7, 8-tetrachloro-dibenzo-paradioxin) which is a side product formed by the hydrolysis of tetrachlorobenzene (cf. Fig. 3.29A). TCDD is found in varying concentrations in raw 2, 4, 5-T, but if there is no subsequent purification process, it is often observed in traces of a few dozen ppm. During the 1960s, commercialized 2, 4, 5-T contained between 0.1 and 100 ppm of TCDD. Since 1970 many countries have compelled manufacturers to improve their production processes in order to minimize the formation of TCDD. In principle, the commercialized 2, 4, 5-T produced in these countries should now contain less than 0.1 ppm of dioxin, although some inspections carried out in Sweden showed that more than 30% of samples analysed contained nearly 1 ppm of TCDD.

TCDD is extremely toxic to vertebrates. It is 1000 times more dangerous than any other dioxin formed by the synthesis of phenoxyacetic acid derivative herbicides. The LD_{50} of TCDD administered orally is between 22 and 45 μg/kg for rats, 10 μg/kg for rabbits and only 0.6 μg/kg for guinea-pigs. Similarly, with cutaneous applications of TCDD the LD_{50} is 0.275 μg/kg for rabbits. As a comparison, this means that TCDD is from dozens to hundreds of times more toxic than strychnine according to the species considered.

In sublethal doses dioxin seriously affects several bodily functions in vertebrates and causes lesions in various organs. Laboratory monkeys and rodents exposed long term to TCDD in their food in doses of 0.5 ppb, have developed generalized leucopoenia (*from* Moore, 1978).

Lipid metabolism is profoundly disrupted and the level of cholesterol in the serum is raised by weak doses of dioxin, while the cholesterol level is lowered in cases of acute poisoning. The phanera can also be affected: loss of hair and nails has been observed in experiments using macaque monkeys. In humans, one of the first symptoms of dioxin poisoning is the appearance of chloracne, a very persistent skin disorder.

One of the target organs of TCDD is the liver, with the plasmic membranes of the parenchyma cells being particularly affected (Truhaut *et al.*, 1974).

Dioxin has recently been confirmed to be a mutagenic substance by the Ames test and by the appearance of chromosomal mutations in a higher plant, *Haenanthus catarinae*, exposed to TCDD, as well as by mutations in rats exposed for 13 weeks to twice-weekly doses of 2 μg/kg.

The teratogenic power of TCDD is also very considerable. When the pregnant females of certain species of mammals were given doses of as little as 0.3 μg/kg, there were abnormalities in their offspring. Moore *et al.* (1973) observed a 55% occurrence of cleft palates and 45% occurrence of renal abnormalities in the offspring of female rats which had been exposed to 3 μg/kg of TCDD given by mouth between the 10th and 13th days of gestation.

Dioxin also has a high embryotoxicity. In the pike (*Esox lucius*), a freshwater Teleostei, embryonic development is inhibited 100% by a concentration of TCDD equal to 10 ppt (Helder, 1980).

Table 3.7 Teratological abnormalities caused by TCDD in rats. (After Moore *et al.*, in Moore, 1978. *Reproduced by permission of Ecological Bulletins, Stockholm*)

| | | Abnormalities observed in the offspring | | | |
| | | Cleft palate | | Renal malformations | |
Days when the pregnant females were treated	Dose of TCDD (in μg/kg)	Number of litters affected	Average number of individuals affected (in %)	Number of litters affected	Average number of individuals affected (in %)
10th–13th	3	12/14	55.4	14/14	95
10th–13th	1	3/16	1.9	15/16	58.9
10th	1	0/18	0	16/18	34.3
10th–13th	0 (control)	0/27	0	0	0

Study of the carcinogenic potential of dioxins generally and TCDD in particular is still too recent for any conclusions to be drawn. However, earlier experiments have shown that TCDD caused hepatomas in rats. The pathological significance of this is much debated, with some experts talking of carcinomimetic effects.

Persistence in the environment—TCDD is a substance with a significant persistence in soils, just like the organochlorine compounds. Although it can be broken down by edaphic micro-organisms, it has a half-life of more than 6 months and even of as much as 3 years, according to some experts.

TCDD can be incorporated into the biomass without seeming to give rise to bioconcentration. In southern Vietnam, the use of 2,4,5-T as a defoliant on mangrove swamps caused the transfer of TCDD into the fish and shellfish in doses of between 18 ppt and 0.81 ppb, depending on the sample.

At the time of the Seveso disaster, the dioxin was analytically detectable in the fats of contaminated domestic animals and in some of their organs. The highest levels found were 1.25 μg/kg in a goat's liver and 0.6 μg/kg in a rabbit's liver. The use of 2,4,5-T to clear the shrubs invading the extensive grazing-lands of the western United States led to dioxin contamination of those regions. However, analyses of the liver and adipose tissue from cattle raised in the polluted zones did not reveal significant traces of dioxin. Only one sample of liver contained detectable concentrations of TCDD while the fats contained between 20 ppt and 60 ppt depending on the sample (*in* Ramel, 1978).

The Seveso disaster, which happened on 10 July 1976, was a striking illustration of the ecotoxicological risks associated with such a toxic and persistant substance as dioxin. At 12:40 p.m. there was an explosion in a reactor producing trichlorophenol at the ICMESA factory belonging to the Swiss Givaudan group and where 350 tonnes/year of trichlorophenol were being manufactured. The reactor had been modified fairly recently and the authorities responsible for the factory had been informed that dioxin was produced in this process. A security valve was opened which sent the contents of the reactor directly into the atmosphere. A cloud of smoke 50 metres high was formed which fell in just a few minutes over an area of 700 metres wide by several kilometres long. On 15 July some domestic animals began to die in suspicious circumstances and on 22 July about thirty people developed skin lesions, a first symptom of contamination. The number of cases of poisoning went up daily at an alarming rate until finally at the end of July it was decided to evacuate the population from the contaminated area.

Three zones were delimited: Zone A, covering about 115 hectares and with 750 inhabitants, was completely evacuated. Zone B, with 5000 inhabitants and covering 250 hectares was put under sanitary control and the 1300 children living there were evacuated for a while during the day to prevent any contact with their contaminated environment. In zone R, covering 1300 hectares and numbering 25 000 inhabitants, measures were limited to prohibiting all agricultural activity (as in zones A and B).

It is estimated that 2 to 3 kg of dioxin fell in zone A and a few dozen grams in zones B and R (*in* Reggiani, 1983). Out of a total of 10 000 people given medical examinations, there were 586 cases of poisoning from Zone A, 392 from zone B and 310 from zone R. Most of the health problems observed were relatively mild and usually dermatological disorders, particularly chloracne. However, some more contaminated individuals suffered hepatic or renal lesions (Hay, 1976). By November 1977, there were still 44 people in hospital.

Nevertheless, it is remarkable that no human death occurred considering the devastating effects of dioxin on livestock in the Seveso region. This was doubtlessly due to the fact that 80% of the dioxin released fell in a part of zone A which was uninhabited and of mainly agricultural use. In addition, the herbivores eating plants whose leaves were found to be covered with 0.1 μg/cm^2 of dioxin in extreme cases, were exposed to much higher doses of the substance than the humans living in the polluted area.

Some eight years after the disaster, the problem of completely decontaminating zone A had still not been resolved. A depth of 35 cm of soil in that zone had been removed entirely and replaced with pure soil. The contaminated soil was piled up in one part of zone A whereas the rest of that zone had been returned to its inhabitants who went back to their homes after the decontamination was complete.

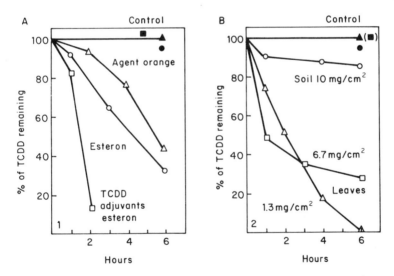

Fig. 3.30 A: Degradation of TCDD as a function of time when a deposit of 2,4,5-T in the form of a defoliant (agent orange or esteron) on a glass slide is exposed to sunlight, and of TCDD added separately to esteron adjuvants. B: Degradation of the TCDD contained in agent orange exposed to the sun after being spread on the surface of the soil or on leaves in different concentrations. *Meaning of the symbols*—Fig. A: △ Agent orange, ○ Esteron, □ Esteron adjuvant and TCDD.
Fig. B: □ agent orange at a dose of 6.7 mg/cm^2, △ agent orange at 1.3 mg/cm^2 on leaves, ○ agent orange at a dose of 10 mg/cm^2 on soil. The same symbols in black denote the control specimens kept in the dark. (After Crosby and Wong, 1977, *Science*, **175**, 1337–1338. *Copyright 1977 by the AAAS*)

The experts were still divided as to how to eliminate the dioxin from the pile of polluted soil. Some suggested that a huge oven should be built capable of burning the soil at a temperature of 1200 °C, at which temperature the TCDD is pyrolysed. It is a paradoxical fact that dioxin, which is easily degraded by sunlight especially when it is in an oily solution, shows considerable stability when it has leached into the soil. Taking account of its almost total insolubility, it is, therefore, not possible to count on dilution to reduce its toxicity with time.

The Seveso disaster cost an estimated £60.5 million, without counting the enormous amount of human suffering, loss of property and other damage which is always difficult to calculate. So far, most of the costs have been borne by the Italian government. Prevention in this case would have been far less expensive, as a simple safety system, costing only a fraction of the investment in the plant, would have been enough to avoid the catastrophe. Furthermore, such a modest expense for a company the size of Givaudan would not have unduly affected the production costs, whose importance to modern manufacturers seems to justify some of the worst attacks on the environment.

Fungicides

Although these are not generally very dangerous to either green plants or animals, they are not without ecotoxicological effects. They apparently play an important part in the growing scarcity of earthworms in agroecosystems, even though other categories of pesticide are just as toxic for these Oligochaeta (*in* Bouché, 1974).

Since the beginning of the 1960s it was noticed that there was a lack of geodrilofauna in orchards treated systematically with Bordeaux mixture. More recently, various researchers have pointed out the sensitivity of earthworms to synthetic fungicides. Captan, thiabendazole, methyl thiophanate and benomyl, when used in orchards at the prescribed dose against spotting agents, cause an increase respectively of 7, 26, 36 and 39 times in the quantity of leaf-litter remaining on top of the soil, due to reduced consumption of the litter by earthworms. The burrowing activity of earthworms is inhibited by a dose of 1200 g/hectare of benomyl. If the same four fungicides mentioned are spread at a rate of 0.78 g/m^2, they cause 100% mortality among a population of Oligochaeta after 18 days of contact (Wright and Stringer, 1973).

In the biomass, earthworms constitute the main group in the soil fauna (an average of 2 tonnes/hectare). As they are so important to the fertility of soils by distributing humus evenly over the upper soil layers, it is clear that the ecotoxicological consequences of their sensitivity to fungicides can no longer be ignored.

Synergism between Pesticides

Pesticides are never used separately and so the simultaneous presence of different pesticides in a terrestrial or aquatic biotype is likely to have synergistic effects.

Just such an effect was shown from a mixture of DDT, parathion and atrazine. In this case, toxicity varied according to the nature of the soil (Liang and Lichtenstein, 1974).

2. Biocoenotic effects

In spite of their importance to applied ecology, the biocoenotic effects of pesticide usage are less well known than the demoecological effects.

Reduction of Food Supply

One of the first ways that pesticides perturb communities is by reducing the plant and animal food supplies available to the species at different trophic levels in agroecosystems (Pimentel and Edwards, 1982).

The gradual disappearance of wild plants from cultivated land due to the systematic use of herbicides constitutes a profound modification of habitat whose consequences are very unfavourable to various species of non-migrant birds living on or near cultivated land. Elimination of wild vegetation deprives these birds of the shelter they need for nest-building or wintering, as well as taking away an indispensable food source for survival during the winter months.

Similarly, the use of non-persistent insecticides (organophosphates and carbamates, for example), while not causing long-term toxicity comparable to that of organochlorines, can still have damaging effects on insectivorous birds: destruction of insects during the reproduction period means that there is no food for the nestlings.

Upsetting the Balance of Nature

Pesticide use can also cause the proliferation of certain plant and animal species by modifying interspecific competition.

The introduction of herbicides into cereal culture eliminated dicotyledonous weeds and resulted in a rapid increase of foxtail grass, an unwelcome species whose spread was favoured by the disappearance of the other weeds.

Although there are other contributory factors, the use of insecticides has caused new species of crop pests to appear, occupying the place left by the almost complete destruction of formerly dominant insects before the crops were treated.

Apart from the elimination of competitors for food supplies, populations dynamic is often even more profoundly modified by the destruction of predators and parasites due to insecticides. For instance, in the United States the excessive use of an organophosphorus insecticide, azodrin, to control cotton plant pests led to a ridiculous situation (after Van den Bosch *in* Huffaker, 1971). Far from reducing the populations of bollworm (*Heliothis zea*) the azodrin eliminated the caterpillars' predators and parasites, which then increased the numbers of pests and damaged bolls, resulting in much greater destruction to the areas of cotton treated with the insecticide than to the untreated control areas.

Effects on the Diversity of Biocoenoses

Herbicides considerably reduce the diversity of biocoenoses. They impoverish plant communities and thereby the zoocoenoses associated with them. The example already quoted of cornfields is a good illustration of this.

Use of 2,4,5-T to clear the undergrowth from large grazing territories or in forestry has similar effects. In the United States, its application over millions of hectares of scrub country covered in sage-brush, *Artemisia tridentata* caused a very noticeable reduction in plant species and in some species of vertebrates associated with those ecosystems, such as the sage-grouse.

In the same way, the variety of microfauna of invertebrates in a limnic environment is necessarily affected by the use of organophosphorus insecticides or carbamates to control parasitic insects. The example was given earlier of the acute sensitivity of Cladocera and copepod crustaceans to these substances.

The diversity of the invertebrate fauna of agroecosystems is also profoundly affected by pesticides. Menhinick (1962) counted 53 species of insects in an area of prairie which had been dusted with insecticide, as against 82 species in a neighbouring control area. The biomass was also significantly less in the treated area.

Effects on the Succession

The succession of animal populations depends closely on that of the phytocoenoses, so herbicides have a greater effect than insecticides on the succession of communities.

Herbicides that are not very selective have a similar effect to fire: they put the ecosystem back to its initial state of colonization by pioneer plants. In some cases, systematic herbicide usage can lead to the formation of a dysclimax. The forests in Vietnam that were completely destroyed by defoliants cannot grow again because the land they once covered has now been taken over by dense bamboo or grasses plantations, making it impossible for reafforestation.

The use of brushwood herbicides to stop ligneous species from colonizing grasslands is blocking the spontaneous evolution of the ecosystem towards a more diversified fruticose shrub stage.

II — Pollution by Chemical Fertilizers

The intensification of agriculture in industrialized countries entails the ever-increasing use of mineral fertilizers. It leads to a treatment of the soil with far greater amounts of chemical fertilizers than the crops can effectively absorb. Misuse of mineral manure results in growing pollution of farmland and continental waters.

The principal fertilizers used in agriculture are potassium salts, nitrates and superphosphates.

1. The superphosphates

These are soluble orthophosphates, non-purified for reasons of production costs. They, therefore, contain traces of numerous toxic metals and metalloids.

Some of these impurities, such as arsenic, are not very mobile in soils and tend to accumulate in the upper layers.

Potassium fertilizers can also contain various impurities. One of these, sylvinite, also has a high proportion of sodium chloride, which has a damaging effect on the colloidal structure of soils.

The impurities contained in mineral fertilizers gradually contaminate farmland with toxic elements which are liable to pass into the edible parts of the plants or to reach phytotoxic concentrations. In the long term, an accumulation of these substances in the upper layer of cultivated soil, where most of the plants' roots are found, constitutes a potential threat to their productivity.

Table 3.8 Principal impurities contained in superphosphates. (After Barrows, 1966.)

Arsenic	2.2 to	12 ppm
Cadmium	50 to	170
Chromium	66 to	243
Cobalt	0 to	9
Copper	4 to	79
Lead	7 to	92
Nickel	7 to	32
Selenium	0 to	4.5
Vanadium	20 to	180
Zinc	50 to	1490

2. The nitrates

The most important cause of pollution of agroecosystems today is the misuse of nitrogen fertilizers. The over-use of nitrates in agriculture results in higher nitrate levels in the tissues of the crops.

Contamination of Plants

Lettuces grown in a soil treated with 600 kg/hectare of nitrates contain 6 g/kg (dry weight) of nitric nitrogen against only 1 g/kg found in lettuces from an untreated control area.

Spinach can easily accumulate nitrates from the soil in their tissues. Schuphan (1965) found up to 3.5 g/kg of nitrates in samples of spinach from Germany. Consumption of spinach grown in soils heavily fertilized with nitrates leads to human health risks because it can cause a type of anaemia—methaemoglobinaemia. This condition develops when there is excessive intake of nitrates or nitrites in food or drinking-water and results from the NO_2 ion combining with the haemoglobin. The substance formed, methaemoglobin, cannot combine with molecular oxygen (from the air). During the industrial processing of any vegetables containing excessive amounts of nitrates, or when they are kept in a refrigerator or while they are being digested, the nitrates can be broken down into very toxic nitrites. Vegetables prepared or preserved in this way can be a potential danger to any children fed on them. There have already been cases in Europe of methaemoglobinaemia due to consumption of spinach contaminated with an excess of nitrates.

Some recent research has shown that NO_2 ions can react with various amines found in the gastrointestinal tract to form carcinogenic nitrosamines.

As a result, experts on the FAO–WHO committee have fixed the maximum level of nitric nitrogen in drinking water at 10 ppm and in food at 75 ppm.

Effects on Surface and Ground Waters

The misuse of nitrogen fertilizers in industrialized agricultural countries also causes serious pollution of ground and surface waters.

Commoner (1970) considered that at least 20% and probably as much as 50% of the nitrates used on the soils of the American cereal-growing regions are not absorbed by the plants but find their way through precipitations into the surface waters. He gave excellent proof of a correlation between the increased use of nitrogen fertilizers in the United States—which had gone up 14 times between 1945 and 1970—and the nitrate level in rivers and streams over the same period. Although this was strongly disputed by the powerful association of American fertilizer manufacturers, Commoner was able to prove irrefutably the role of excessive nitrogen fertilizer use in the pollution of continental waters. He based his evidence on the fact that the isotopic proportion of heavy nitrogen (^{15}N) to normal nitrogen (^{14}N) is not the same in natural nitrates, which are produced by various nitrifying micro-organisms, as it is in synthetic nitrates, produced using atmospheric nitrogen.

The following isotopic *coefficient* can be defined:

$$\delta_{15_N} = \left(\frac{\frac{[^{15}N]}{[^{14}N]} \text{ sample}}{\frac{[^{15}N]}{[^{14}N]} \text{ atmosphere}} - 1 \right) . 10^3$$

This coefficient is less than or equal to 2‰ in the case of synthetic nitrates, whereas it is over 10°/00 for natural nitrates produced by bacterial action on organic matter. By knowing the $\delta^{15}N$ of nitrates in an area of continental water, it is possible to calculate quite precisely how much the water is polluted by nitrogen fertilizers.

Mariotti *et al.* (1975), using the same method, showed that the surface waters of the Brie region, an area of extensive cereal cultivation, were highly contaminated by nitrates from fertilizers at certain times in the year when

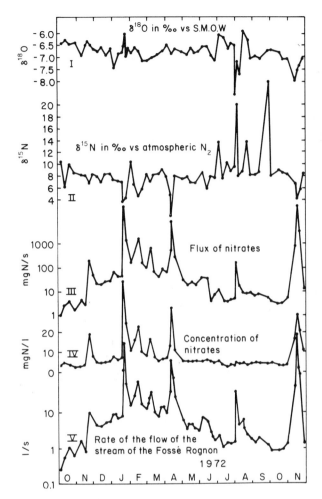

Fig. 3.31 Experimental proof of the contamination of surface waters by nitrates used on the cereal crops in the Brie region. These diagrams show a considerable decrease in $\delta^{15}N$ between January and June due to leaching of nitrogen fertilizers from the fields. (After Mariotti *et al.*, 1975. *Reproduced by permission of Paul Germain*)

leaching of over-fertilized soils occurred. The study of a complete hydrological cycle by these researchers revealed that the balance of water flowing out of the basin contained a high proportion of nitrates with a low $\delta^{15}N$ (artificial fertilizers), which led them to conclude that there was excessive use of nitrogen fertilizers in the area studied.

It is worth pointing out that most of the wells which supply drinking water to rural communities in the Paris basin are situated in cereal-growing areas where present levels of nitrates are over the 10 ppm maximum allowed by the WHO.

B—ATMOSPHERIC POLLUTANTS

The principal atmospheric pollutants can be divided into two major groups: the gases and the solid particles, often wrongly called 'aerosols'. The gases constitute 90% of airborne pollutants and the solid particles make up the remaining 10%.

I—Combustion Products

The burning of various forms of fossil carbon plays an important part in atmospheric pollution.

1. Carbon monoxide

Sources

Although carbon monoxide is found naturally in the atmosphere in a very low concentration (average of 0.1 to 0.2 ppm), it is one of the main pollutants produced by automobile exhaust. In 1970 the amount of carbon monoxide emitted into the atmosphere in the United States alone was estimated at some 102 million tonnes.

Table 3.9 shows estimates of the total quantities of carbon monoxide released into the atmosphere from natural sources and human technology. According to this table, it seems that the amounts produced by man (257×10^6 t) are clearly greater than those produced by natural biogeochemical processes. However, there are still significant doubts relating to the nature and importance of the natural sources of carbon monoxide.

Table 3.9 Estimation of the principal sources of carbon monoxide on a global scale. (After Jaffe 1970 *et al.*, *in* Singer, 1970)

	Sources	Consumption in 10^6 tonnes/year	CO emissions in 10^6 tonnes/year
natural	Ocean		9
	Oxidation of plant terpenes		61
	Volcanism		?
	Total		> 70
technological	Petrol	379	193
	Coal	3074	12
	Wood (combustible)	466	16
	Incineration of refuse	500	25
	Forest fires	7*	11
	Total		257

* Area burnt annually in 10^6 hectares (in 1968).

Weinstock and Niki (1972) in a study of a mathematical model of the stratosphere, considered that significant quantities of carbon monoxide were forming at this level due to photodissociation of the methane present in the atmosphere and its subsequent transformation into carbon monoxide.

Various living beings can assimilate carbon monoxide. The air bladders of algae of the genus *Neocystis* contain 800 ppm of this gas and it can also be found in significant amounts in the pneumatophores which sustain the *Physalia* and other colonial Cnidaria of the siphonophoran group. Weinstock and Niki calculated finally that the quantity of carbon monoxide produced by the different biogeochemical processes could be fifteen times greater than that produced by man by the end of the 1960s.

Nevertheless, it is a fact that the highest concentrations of carbon monoxide are found in the polluted atmosphere above urban areas. For large cities the average amount is about 20 ppm with peaks of more than 100 ppm during traffic jams.

Considering the quantities of carbon monoxide released by human activity, an increase of about 0.04 ppm/year could be expected in the concentration of carbon monoxide in the atmosphere. However, the concentration level is stable in remote corners of the globe, showing that biogeochemical mechanisms exist which degrade this gas and transform it into other compounds.

In the stratosphere, the presence of free OH–radicals is a significant factor in neutralization of carbon monoxide, according to the reaction:

$$CO + OH^- \rightarrow CO_2 + H^+$$

In the soil, certain bacteria have an important role in neutralization processes. *Bacillus oligocarbophilus* and *Clostridium welchii* are both capable of oxidizing carbon monoxide to carbon dioxide. *Methanosarcina barkerii* and *Bacterium formicum* convert carbon monoxide either to methane or to carbon dioxide depending on whether there is a hydrogen donor present or not (Inman *et al.*, 1971). The bacterial flora of the soil is, therefore, an essential element in the carbon monoxide cycle.

Effects on Organisms

Carbon monoxide can have different effects on living beings according to their taxonomic position. In ordinary concentrations it is harmless to plants, but beyond that it interferes with nitrate metabolism and in high concentrations it is clearly phytotoxic, inhibiting respiratory processes in particular.

Warm-blooded vertebrates are very sensitive to carbon monoxide as it combines irreversibly with haemoglobin. The carboxyhaemoglobin formed in this way is incapable of fixing oxygen and transferring it to the cells, causing asphyxiation in victims.

Inhalation of air polluted with 6.4×10^3 ppm of carbon monoxide causes headaches and dizziness within 2 minutes and unconsciousness and risk of death within 15 minutes. A concentration of 100 ppm represents the maximum 'permissible' limit.

2. Hydrocarbons

The principal sources of atmospheric pollution by hydrocarbons are the internal combustion engine, domestic and industrial uses and accidental leaking of petroleum products and motor fuels. The most significant of these sources is the incomplete combustion of fuel in petrol engines and oil-fired boilers. Also during such combustion processes dangerous carcinogenic polycyclic hydrocarbons are produced: benzo-3,4-pyrene, benzanthracene, fluoranthrene, cholanthrene, etc. These substances are found in gaseous or particulate form (soot and tar) in the waste from combustion products from homes and industry, diesel engine exhaust (cf. for example, the unacceptable volume of fumes from hill-climbing lorries) and to a lesser extent in the exhaust from petrol engines.

Vegetation and bacterial fermentation can also give off hydrocarbons, mainly terpene and methane. The bluish haze seen above forest areas in fine weather can be attributed to the release of terpene from the trees. A total of some 100×10^6 tonnes per year of hydrocarbons are released into the atmosphere by vegetation and bacteria.

3. Nitrogen derivatives

Nitrogen Oxides

These are secondary products from combustion reactions of various carbon derivatives with air. Although both nitric oxide (NO) and nitrogen dioxide (NO_2) are natural constituents of the atmosphere, they are produced in large quantities during combustion at high temperature and under high pressure, two conditions found in the internal combustion engine.

Nitrous oxide (N_2O) is the most abundant nitrogen oxide in a non-polluted atmosphere, with an average concentration of 0.25 ppm. It does not contribute to atmospheric pollution from human activity, in contrast to NO_2 which is a significant factor in this type of pollution.

NO is transformed spontaneously into NO_2 above a temperature of 600 °C through the reaction:

$$2NO + O_2 \rightarrow 2NO_2 + 28.4 \text{ kcal} \tag{1}$$

As motor vehicle automobile exhaust contains an average of 1000 ppm of NO, it is easy to understand the importance of urban traffic in atmospheric pollution from nitrogen oxides.

However, the burning of fuel-oils and coal is an even more important factor in the production of these pollutions (cf. Table 3.10).

Table 3.10 Principal sources of emission of nitrogen derivatives into the atmosphere. (*In* Varney and MacCormac, 1971. *Reproduced by permission of D. Reidel Publishing Co.*)

Compounds	Source	Quantity in 10^6 t/year (in nitrogen equivalent)
N_2O	Bacteria	340
NO	Bacteria	210
	Motor vehicles	2.0
	Combustion of coal	7.4
NO_2	Combustion of fuel-oils	4.5
	Other combustion	0.7
NH_3	Bacteria	860
	Combustion	3.1
	Bacteria	860
NH_3		
	Combustion	3.1

Emissions of NO_2 into the atmosphere from human activity amount in fact to more than 14×10^6 tonnes/year. (Expressed in nitrogen-equivalent. According to Söderlund and Svensson (1976), the anthropogenic emissions of NO_x could be between 8×10^6 tonnes and 25×10^6 tonnes of nitrogen-equivalent per year.)

Nitrogen dioxide does not remain in the atmosphere for long. After a few days it is converted there into nitric acid and then into nitrates by reacting with various cations, particularly NH_3. The ammonium nitrate thus formed produces inframicroscopic particles which are then quickly carried down to the soil by precipitations.

Peroxy-acyl-nitrates (PAN)

These occur in polluted urban atmospheres subjected to strong sunlight, conditions which cause the genesis of oxidizing photochemical smogs (in Los Angeles, for example).

The PAN are formed by the following general reaction:

$$R-\underset{\underset{O}{\|}}{C}-O-O+NO_2 \rightarrow R-\underset{\underset{O}{\|}}{C}-O-O-NO_2 \qquad (2)$$

where R is an aliphatic, aromatic or heterocyclic radical.

The main members of the aliphatic series are:
— peroxy-acetyl-nitrate ($R = CH_3 -$)
— peroxy-propionyl-nitrate ($R = CH_3 - CH_2 -$)
— peroxybutyryl-nitrate ($R = CH_3 - CH_2 - CH_2 -$)

The PAN occur in oxidizing smogs following a series of photochemical reactions involving unburnt residue hydrocarbons from incomplete combustion processes, ozone and nitrogen oxides.

Ozone is itself an atmospheric pollutant formed in a reaction between oxygen and NO_2. In the presence of ultraviolet rays,

$$NO_2 + O_2 \xrightarrow{UV} NO + O_3 \tag{3}$$

The ozone will then react with the hydrocarbons in the atmosphere which it converts into alkylperoxides and then into peroxy-acyls $(R - C(O) - O - O -)$. These peroxy-acyls will then combine with NO_2 to form the PAN as in equation (2).

In the polluted atmosphere of southern California, up to 2.65 ppm of nitrogen oxides and 58 ppb of PAN have been observed.

Effects on Living Beings

Nitrogen derivative gases have a toxic effect on most terrestrial organisms.

Of the nitrogen oxides, only NO_2 is truly phytotoxic, although this is only from concentrations of at least 0.5 ppm, which very rarely occur. In addition, the toxicity of NO_2 to plants is stronger in weak light, as the processes that reduce the nitrates are slower when the plants are inactive.

The PAN are far more phytotoxic. Solanaceae and Compositae are very sensitive to these substances and develop severe foliar lesions after only 4 hours' exposure to cencentrations of less than 15 ppb. These lesions are characteristic of PAN contamination, which also causes the lamina to turn a silvery and then bronze colour as a result of the vacuolization of the cells in the mesophyll of the leaf. In higher concentrations, the PAN cause necrosis of the foliar parenchyma arranged in transversal bands, especially in monocotyledons. The PAN inhibit photosynthesis and, therefore, slow down the growth of plants exposed to weak concentrations of the substances.

The various nitrogen derivative gases present in polluted atmospheres are also toxic to animals. In high concentrations, above 10 ppm, NO_2 causes a hyperplasia of the pneumocytes and in the long term it destroys the bronchioles. Nevertheless, the effects of permanent exposure to concentrations of less than 1 ppm are not yet fully known, although it is suspected that they play a part in certain forms of chronic bronchitis.

The PAN and their aromatic homologues, peroxybenzoyl nitrates, cause serious eye irritation (watering, conjunctivitis) even in concentrations of as little as 10 ppb. Some of these compounds can induce eye-watering in doses 100 times weaker than those needed by formol. As they are regularly found in amounts of over 50 ppb in some California cities, particularly Los Angeles, the eye irritation caused on days of heavy pollution can lead the authorities to be obliged to close schools.

4. Sulphur dioxide (SO_2)

This has become a major atmospheric pollutant.

Sources

Under natural conditions, sulphur dioxide is found in the atmosphere in minute traces: its average concentration in the troposphere has been estimated at 0.2 ppb.

The major source of SO_2 pollution of the atmosphere is the burning of fossil fuels. In 1970 the amount of SO_2 given off from the combustion of various fossil carbon derivatives was calculated to be some 145×10^6 tonnes. Of this total, coal represents about 70% and fuel-oils 16%. Coal can in fact contain over 5% sulphur and industrial heavy fuel-oils at least 3% sulphur.

Recent figures for the emission of sulphur equivalents into the atmosphere by man are 100×10^6 tonnes per year, a vast amount if one realizes that natural sources of atmospheric sulphur from the continents and oceans would only come to 40×10^6 tonnes per year. It shows, therefore, how human activity has profoundly disrupted the biogeochemical cycle of this element.

Urban atmospheres are permanently contaminated by sulphur dioxide at the present time. During the 1960s, the average SO_2 content of the atmosphere was 0.17 ppm in the Paris region. However, after the oil crisis in 1973 energy conservation measures and reduction in consumer levels reduced this average below 0.1 ppm by the beginning of the 1980s and to 0.06 ppm in 1985.

The record sulphur dioxide concentration is held by Los Angeles with peaks of 2.49 ppm (cf. Table 3.11).

Sulphur dioxide does not remain in the atmosphere *ad infinitum*. It undergoes various transformations which form part of the normal biogeochemical cycle

Table 3.11 Contamination levels for the main atmospheric pollutants in the Los Angeles urban area. (After Stern *in* Simmons, 1974. *Reproduced by permission of Century Hutchinson Ltd.*)

Contaminants	Usual range of concentration (in ppm)		Maxima observed (in ppm)
	Days of smog	Normal days	
Sulphur dioxide	0.15– 0.70	0.15– 0.70	2.49
Aldehydes	0.05– 0.60	0.05– 0.60	1.87
Carbon monoxide	8.00–60.00	5.00–50.00	72.00
Hydrocarbons	0.20– 2.00	0.10– 2.00	4.66
Nitrogen oxides ($NO + NO_2$)	0.25– 2.00	0.05– 1.30	2.65
Oxidizers	0.20– 0.65	0.10– 0.35	0.75
Ozone	0.20– 0.65	0.05– 0.30	0.90

of sulphur. When it comes into contact with atmospheric water vapour, the SO_2 is turned into sulphurous acid by the reaction:

$$SO_2 + H_2O \rightarrow H_2SO_3 + 18 \text{ kcal} \qquad (4)$$

Then, in the presence of ultra-violet rays, the sulphurous anhydride reacts with the oxygen in the air to form sulphur trioxide (SO_3):

$$SO_2 + \tfrac{1}{2}O^2 \xrightarrow[-x]{h\nu} SO_3 + 22 \text{ kcal} \qquad (5)$$

Finally, in a polluted atmosphere the following reaction takes place:

$$SO_2 + NO_2 + H_2O \rightarrow H_2SO_4 + NO \qquad (6)$$

Eventually, the SO_3 and SO_3H_2 formed in reactions (5) and (4) will be transformed spontaneously into sulphuric acid.

Pollution by SO_2 and Acid Precipitations

(a) Acid rains—The combination of SO_2 with atmospheric oxygen and water vapour will eventually result in the formation of sulphuric acid. This is a very hygroscopic substance and forms toxic fogs, helping to create the acid smogs over cities situated in humid temperate climates.

The average retention time for SO_2 in the troposphere is very short, from 2 to 4 days. It is quickly transformed to SO_4H_2 which, due to its strong affinity for water, is soon carried in precipitations to the soil surface.

Oden (1968) quoted in Likens *et al.* (1979) was able to cite SO_2 pollution as the cause of the continual decrease in pH observed in precipitations over the whole of western Europe since the middle of the 1950s. More recently, Likens and Bormann (1974) and Likens *et al.* (1979) pointed out the higher acidity of both snow and rainfall over the entire north-east of the United States. They also noted a tendency for the pH of the precipitations to decrease over the past 20 years. The pH is down to 4 in the forests of New Hampshire, situated 1000 km away from heavily industrialized zones. It is even lower throughout New York State where these researchers found that every year some rain had a pH of as low as 3, with a lowest level of 2.1 recorded in precipitations in November 1964.

In Europe, the highest level of acidity ever observed in precipitations was recorded in Pitlochry, Scotland on 10 April 1974 when during a storm the pH of the rainwater reached 2.4. The world record for acid rains is currently held by the town of Wheeling, West Virginia, USA where in 1979 the pH of precipitations was 1.7, which is somewhere between the acidity of lemon juice and car battery acid (according to Vie le Sage, 1982). The study of snow sampled

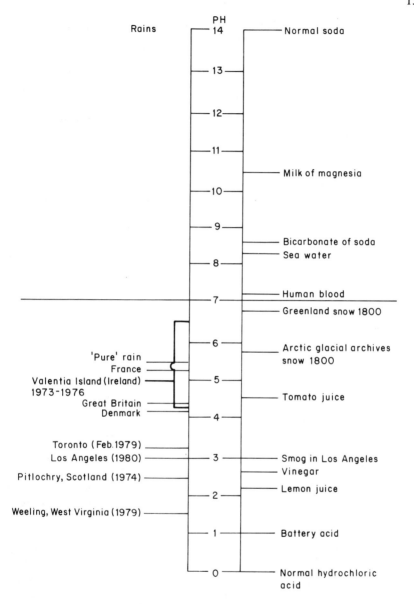

Fig. 3.32 A pH scale showing the increase in the acidity of rainwater and the highest levels recorded. (After Vie le Sage, 1982)

in arctic zones and in alpine glaciers confirms the increasing acidification of precipitations during the industrial era. For instance, the pH of snow in Greenland 180 years ago was between 6 and 7.6 with an average of 6.8, which is very close to being neutral (Fig. 3.32). In contrast, the lowest annual average (3.78) was recorded in De Bilt, Holland (Likens et al., 1979).

Among the anions contained in acid rain, SO_4 alone accounts for 70% of the total, with the remaining 30% consisting largely of the anion NO_3^-.

The levels observed in the north-east of the United States are, therefore, comparable with those found by Oden (1968) quoted in Likens et al. (1979) in southern Sweden. In this region, specialists estimate that 70% of the sulphur present in the atmosphere comes from distant industrialized zones: the Ruhr and Great Britain (cf. Ch. 2, page 76). In certain remote regions of Scandinavia, the concentration of H^+ ions in the precipitations has gone up by more than 200 times since 1956.

(b) Ecological consequences—Although not entirely understood as yet, the ecological effects of acid rains are both varied and complex. They can be catastrophic in limnic environments, particularly in oligotrophic lakes situated on crystalline rock and whose water supply comes more from precipitations than from streams.

Acid rains seem to be the sole cause of the continual decrease in the pH of a large number of Scandinavian lakes, as observed between 1965 and 1970 (Oden and Ahl, 1970, *in* Oden, 1976).

During the 1960s, many Canadian lakes situated in the path of dominant winds carrying SO_2 from important metallurgical plants experienced a lowering of the pH equal to 100 times its original level. In 1970, more than 30 of these lakes had a pH of 4.5 or less, which was considered a critical level by the author of the report (Beamish, 1974).

Schofield (1965) gave an earlier example with the case of an oligotrophic lake in the Adirondack mountains, USA, whose total alkalinity reached from 12.5 to 20 mg/l (expressed in equivalent CO_3Ca) with a pH of 6.6 to 7.2 in 1938. By 1960 the alkalinity was less than 3 mg/l and the pH had dropped to between 3.9 and 5.8, depending on the time of year. In the Adirondack mountains in 1930 only 8 out of the 214 lakes had a pH of less than 5, but by 1974 this number had gone up to 109, of which 39 were so acidified that they had almost become azoic.

In Ontario, Canada, the Salmonidae and bass (*Micropterus* sp. Centrarchidae) have disappeared over the last decade from more than 2000 lakes out of the total of 4000 lakes in the State.

The consequences of acidification of lakes by SO_4H_2 carried in precipitations are that biological productivity of the lakes is considerably reduced. A lower pH causes a sharp decrease in the diversity and primary productivity of the phytoplankton (Fig. 3.33). It, therefore, indirectly affects secondary production by lessening the amount of food available to the animal consumers situated on higher trophic levels (Almer et al., 1974).

This acidification also has a direct toxic action on freshwater fish, as shown by the high mortality, particularly among young salmonids in Canadian and Swedish lakes.

Effects of Sulphur Dioxide on the Biocoenoses

SO_2 has a large number of different ecotoxicological effects with serious results

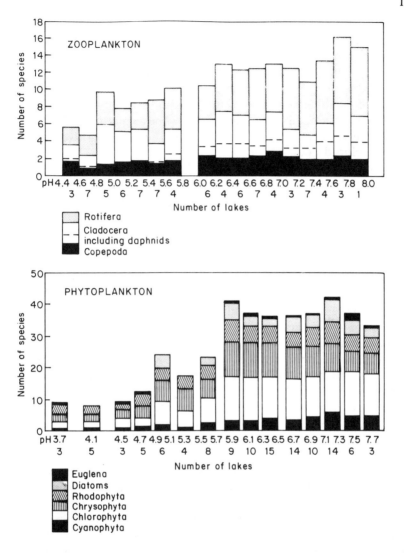

Fig. 3.33 Effects of the acidification of Swedish lakes by acid atmospheric pollutants (particularly SO_2) on the diversity of their planktonic population. There is a significant drop in the number of species of phytoplankton and zooplankton below pH 6. (From *Ambio*, 1974, vol. III, No. 1, Almer *et al. Reproduced by permission of the Royal Swedish Academy of Sciences*)

on both plants and animals. Its toxicity to plants has been known for some time. At the end of the eighteenth century in Sicily it was blamed for the disappearance of all the vegetation round the Calcaroni, primitive sulphur-refining plants.

(a) Phanerogams—These cannot grow normally in an atmosphere where the average annual sulphur dioxide content is between 10 and 18 ppb (Nash, 1973).

Even in the most resistant plants, foliar cankers appear after only 8 hours' exposure to relatively weak doses of 0.25 ppm. The SO_2 penetrates the plant primarily through the stomata and damages the cells of the mesophyll with its reducing properties.

In **dicotyledons**, signs of chronic contamination are characterized by bifacial necrosis of the foliar lamina along the veins and extending towards the petiole. Necrosis can also occur in the apical region. The colour of the damaged tissues varies from an ivory tint to reddish-brown. The most sensitive leaves are the newly-developed ones.

In **monocotyledons** it causes an apical necrosis which gradually extends to the base of the lamina along the veins. The damage also starts at the end of conifer needles and has an orangy-red colour. With the chronic exposure of conifers, the foliar lamina becomes chlorotic and then damage appears in the parenchyma often in transversal bands.

A microscopic examination of leaves that have suffered chronic SO_2 contamination shows a flaccid degeneration of the parenchymatous cells accompanied by a lysis of the chloroplasts whose chlorophyll is then released into the whole of the cytoplasm.

Even long-term exposure to low concentrations of SO_2 can cause a clear reduction in metabolic activity.

A large number of studies on the ecotoxicological consequences for plants of SO_2 contamination have been carried out over the last 50 to 60 years, due to the economic importance of this question. In 1922 O'Gara (*in* Linzon, 1971) established by fumigation experiments a classification of about 100 species of phanerogams in descending order of sensitivity. Various other research into the same subject has shown that the sensitivity of phanerogams to SO_2 differs considerably with their taxonomic position and even sometimes within the same taxonomic group.

Some of the most sensitive vascular plants to SO_2 are certain conifers: foliar lesions appear on the *Pinus strobus* after only 8 hours' exposure to 0.1 ppm. In the Ruhr, Knabe (1971) observed that the pines could not survive an average annual concentration level of 80 ppb and that damage occurred with as little as 20 ppb. Some Swedish botanists have shown that the growth of coniferous forests can be inhibited by average doses of 20 ppb, which they considered to be the maximum permissible concentration. More recent studies have provided evidence that pine forests in an area of Finland where atmospheric SO_2 pollution oscillates between 14 ppb and 19 ppb develop various physiological disorders (Havas and Huttenen, 1980). The needles of these conifers contain an average of 1.35% of sulphur compared to their dry weight (between the two extremes of 1.1 and 2.6) as against a level of 0.7 to 0.9% in non-polluted zones. There is a clear correlation between the level of sulphur contamination of the needles and their low water-content. It would seem that the slight damage done by the relatively weak concentrations of SO_2 to the functioning of the stomata causes greater transpiration from the conifer needles. Examination under an electron microscope has also shown that the mesophyll of apparently healthy leaves has damage to the thylakoids in their chloroplasts, even with very low levels of SO_2.

As a general rule, pines are considerably more sensitive than spruces and deciduous trees more tolerant than conifers.

Herbaceous plants, cultivated Leguminosae, Compositae (chicory, lettuces, for example) and the cotton-plant are all very sensitive to SO_2, followed by crucifers and most cereals, although maize and asparagus are very tolerant. The weeds that grow in or near crops are also on the whole very sensitive to SO_2, but some of them such as the plantain (*Plantago lanceolata*), starwort (*Stellaria media*) and pimpernel (*Anagallis arvensis*) are very resistant.

Considerable damage can be caused by SO_2. There have been some well-published court cases involving damage done to crops situated near certain industries that release large quantities of SO_2 into the atmosphere.

Forests can be particularly hard-hit by SO_2 pollution. It is estimated that in the EEC alone nearly 500 000 hectares of forest are threatened with total devastation between now and 1990 due to sulphur dioxide emissions from petrochemical plants and the burning of heavy industrial fuels. In West Germany—the EEC country which is second only to Great Britain in releasing the greatest amount of sulphur into the atmosphere—the combined effects of SO_2 gas and acid rains are threatening their most important coniferous forests. In the Bavarian forests which cover more than 80 000 hectares, the first signs of damage to the fir-trees (*Abies pectinata*) and spruces (*Picea excelsa*) appeared in 1979. The trees most seriously affected are the oldest ones whose needles die and fall *en masse*. In addition, the germination rates of the conifer seeds have almost come to a standstill under the effects of acid precipitations. In the Black Forest (Schwarzwald), 64 000 hectares of conifers are showing signs of damage with the number of spruces rapidly diminishing (according to Lubinska, 1982).

Without question, it is in North America where the most spectacular cases of damage to vegetation by SO_2 have occurred. There is a large metallurgical plant in Ontario, Canada, which has been treating pyritic cupro-nickels since 1888. During the 1950s the plant was emitting about 2×10^6 tonnes of SO_2 into the atmosphere per year. As a result, some 200 000 hectares of pine forest surrounding the factory have been virtually annihilated (Linzon, 1971).

(b) Cryptogams—The scarcity of lichens in towns and cities is a recognized phenomenon since the middle of the nineteenth century. On 13 July 1866, Nylander gave a paper to the Botanical Society of France on the lichens in the Luxembourg Gardens. In the report of the discussions which followed are remarks made by the botanist Cosson who confirmed that the decreasing number of lichens in cities was a result of the smoke and gases produced which made the atmosphere unfit for them to grow in.

Since then much research has been devoted to the effects of atmospheric pollution on these cryptogams. It has proved the dominant role of SO_2 in the scarcity or absence of lichens in inner city areas, although various other

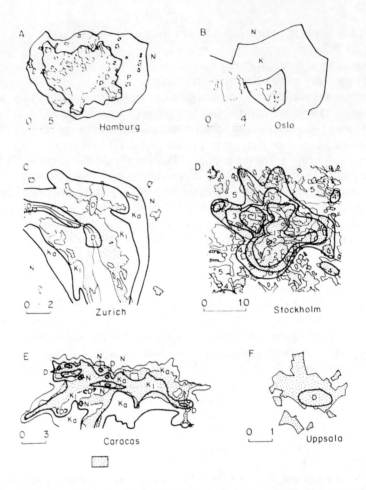

Fig. 3.34 Examples of the zonation of lichens in cities. D = no lichens present, K = the 'struggle zone', or a zone where the structure of the lichen biocoenosis is profoundly changed by the pollution (K_1 = very rare lichens, K_a = more abundant lichens but with little diversity), P = zone of community reconstruction, N = normal zone. In Fig. D, there is the following gradation: 1 = D, 2 = K_1, 3 = K_a, 4 = P, 5 = N. (After various authors *in* Laundon, 1973. © *The University of London. By permission of the Athlone Press*)

factors can also play a part such as other atmospheric pollutants or ecological conditions peculiar to the urban environment, like the relatively low humidity. However, it is a fact that no lichens can survive an average annual SO_2 concentration of more than 35 ppb. They are, therefore, excellent bioindicators of atmospheric pollution.

Numerous zonation studies have been made on lichen communities since 1930, particularly in Scandinavia, Great Britain and central Europe. They show the lichens distributed in approximately concentric zones. The central and most polluted area of cities has no lichens; then there is an intermediate zone where the structure of the lichen community is profoundly changed (often called the 'struggle zone'); and lastly a peripheral zone where the lichen population is normal. There also appears to be an excellent correlation between the level of SO_2 pollution and the reduction in lichen population diversity (cf. for example, Skye, 1968). For each species there exists an annual average concentration limit above which it cannot survive. These tolerance limits are so strict that Hawksworth (1973) published a scale which enabled the average winter concentration of SO_2 to be calculated using lichen surveys.

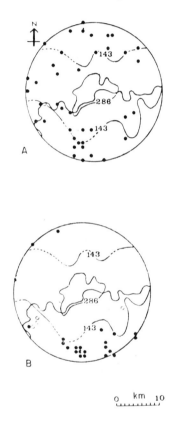

Fig. 3.35 Correlation between the average amount of SO_2 in the atmosphere (expressed in μg/m³) and the distribution of the lichens *Calopalca heppiana* (A) and *Xanthoria parietina* (B) in the London area. ○ : surveys before 1950, ● : surveys after 1960. (After Laundon, 1973. ©*The University of London 1973. By permission of the Athlone Press*)

A crustose lichen, *Lecanora conizaeoides*, seems to be particularly toxicotolerant to SO_2. Although this species was not present in Great Britain in the middle of the last century, it spread widely throughout urban agglomerations from 1870 onwards (Laundon, 1973). As a general rule, it is the crustose lichens which have the highest level of tolerance to SO_2, followed by foliose lichens and then fruticose lichens.

Table 3.12 Qualitative scale for estimating the average winter concentration of sulphur dioxide in the atmosphere using a survey of the structure of the population of epiphytic lichens (in England). (*In* Hawksworth, 1973. © *The University of London. By permission of the Athlone Press*)

Zones	Species composition	Average winter concentration of SO_2 (in $\mu g/m^3$)[1]
0, 1	No lichens	< 170
2	*Lecanora conizaeoides*	~ 150
3	*Lecanora conizaeoides, Lepraria incana*	125
4	*Hypogymnia physodes, Parmelia saxatilis, Lecidea scalaris, Lecanora expallens*	70
5	*H. physodes, P. saxatilis, P. glabratula, P. subrudecta, Parmeliopsis ambigua, Lecanora chlorotera*	
	Ramalina farinacea and *Evernia prunastri* appear	60
6	*Parmelia caperata, P. tiliacea, P. exasperatula, Pertusaria numerous, Graphis elegans*	50
7	*P. caperata* and *P. revoluta, P. tiliacea, P. exasperatula*	
	Usnea subfloridana, Pertusaria hemispherica Rinodina roboris appear	40
8	*Usnea ceratina, Parmelia perlata, P. reticulata, Rinodina roboris. Usnea rubiginea* quite frequent	35
9	*Lobaria pulmonaria, L. amplissima, Pachyphiale cornea, Dimerella lutea, Usnea florida*	30
10	*Lobaria amplissima, L. strobiculata, Sticta limbata, Pannaria* sp., *Usnea articulata U filipendula*	normal (= pure air)

[1] (1 ppm = 2860 $\mu g/m^3$ of air)

Following on from this, some researchers have proposed empirical formulae which can be used as a scale for evaluating atmospheric purity (Index of atmospheric purity = IAP). The index is calculated from phytocoenotic surveys revealing the species diversity in the community, Iserentant and Margot (1964) *in* Ferry *et al.* (1973) define the following IAP:

$$IAP = \frac{n}{100} (\Sigma_1^n Q \times f)$$

where n = number of lichen species in the area studied
f = frequency of each species
Q = the index of toxiphobia for each species.

With the growth in SO_2 pollution which accompanies the anarchic urbanization and industrialization so characteristic of most developed countries,

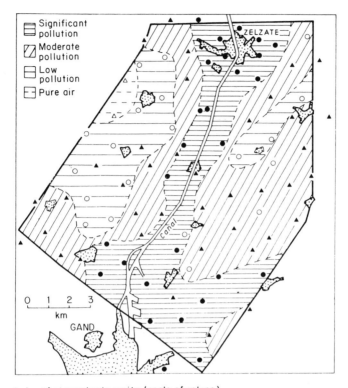

Fig. 3.36 Map showing levels of atmospheric pollution in the Gand-Zelzate area of Belgium, established using the IAP method of Iserentant and Margot. (After Sloover *in* Ferry *et al.*, 1973. ©*The University of London. By permission of the Athlone Press*)

there has been a significant decline over the past decades in lichen phytocoenoses: in England, for example, some 87 species have disappeared since 1960 in areas where SO_2 levels are equal to or above 65 $\mu g/m^3$ or 27 ppb during the winter months.

Quite recently there have been experiments to replant lichens in different zones of Danish towns from where the lichens had disappeared owing to higher levels of air pollution. The species chosen, *Hypogymnia physodes*, is moderately tolerant of SO_2. In this way it has been possible to evaluate *in situ* the effects of increasing SO_2 concentration in the atmosphere over Copenhagen on the thalli of the lichen, as a function of the extent of microscopically observable lesions (*in* Johnsen, 1980).

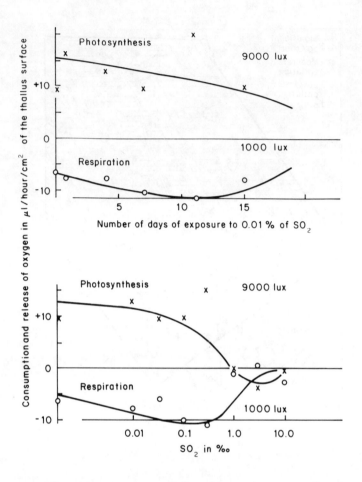

Fig. 3.37 Apparent photosynthetic activity on discs of lichen exposed to various concentrations of SO_2 (upper graphs) and the respiratory activity of the same fragments of thalli (lower graphs) measured as a function of the length of exposure (diagram A) and as a function of the SO_2 concentration (diagram B). (After Pearson *in* Ferry *et al.*, 1973. ©*The University of London 1973. By permission of the Athlone Press*)

The Action of SO_2 on Plants

Sulphur dioxide causes serious lesions of the photosynthetic organelles both in phanerogams and cryptogams.

Rao and Leblanc (1965) demonstrated that lichens exposed to 5 ppm of SO_2 developed cellular plasmolysis and there was a change in the chloroplasts of the symbiotic alga (*Trebouxia*) and black spots appeared in the part of the thallus occupied by the alga. The researchers were able to isolate sulphuric acid and magnesium ions in the acetonic extracts from the thalli treated with SO_2 and from the phaeophytin in the ether-soluble fraction. From this they deduced that the gas had used its reducing properties to break down the chlorophyll according to the reaction:

$$2H^+ + \text{chlorophyll a} \rightarrow Mg + \text{phaeophytin a}$$

The brown spots seen in the gonidial cells of the lichen could come from the accumulation of phaeophytin a.

Pearson and Skye (1965) cited in Skye (1968) have also shown the inhibiting effects of SO_2 on photosynthesis and its damaging action on respiration in lichens exposed to it in varying concentrations. Other research on phanerogams confirms the influence of SO_2 on chlorophyllian assimilation. It has been shown to act by disrupting the Calvin cycle. Dicotyledons grown in the presence of $^{35}SO_2$ synthesize cysteine marked with ^{35}S, proving that the gas interferes with the dark phase of photosynthesis.

The Action of Sulphur Dioxide on Animals

It is extremely toxic to mammals. The thresholds of acute toxicity are only rarely reached in the case of an accident, as it has a characteristic pungent odour even in concentrations as low as 0.5 ppm.

Nevertheless, the problem of how to evaluate its long-term toxicity has become important, given its increased levels in the urban atmosphere.

It has been shown that at a concentration of 0.2 ppm, sulphur dioxide causes conditioned reflexes of the vegetative nervous system in the cerebral cortex, and that at 1 ppm it can diminish pulmonary elasticity in hypersensitive subjects (*in* Masters, 1971). Laboratory rodents raised permanently in an atmosphere containing 2 ppm of SO_2 show signs after a few days of a PAS-positive hypersecretion from their bronchial mucous cells.

Various epidemiological studies have also shown the importance of sulphur dioxide in the genesis of chronic bronchitis and emphysema although its role in these conditions is considerably less than cigarette-smoking, the curse of so-called developed countries.

Research carried out in the United States (Sheppard *et al.*, 1980) proved that asthmatics are hypersensitive to SO_2 in concentrations far below those previously considered harmful. As a result, there has been a dispute between the *EPA* and the industries concerned because the new Clean Air Act could be used to lower quite significantly the present standard for SO_2 levels in the atmosphere over urban zones (Smith, 1981).

5. Ozone

Ozone is a natural constituent of the atmosphere. It is not distributed evenly: it is found at a maximum density in the stratosphere between the altitudes of 20 and 30 km, depending on the latitude (cf. Ch. 2, page 74).

The concentrations of ozone found at different altitudes are the result of a very complex dynamic balance involving a large number of reactions which produce and destroy the ozone.

Sources

Ozone is formed from atmospheric oxygen which though normally in its molecular state (O_2), can be broken down by ultraviolet radiation whose wavelength is between 100 and 2450 Å:

$$O_2 + h\nu \rightarrow O + O \tag{7}$$

The atomic oxygen produced in this way will react with the molecular oxygen to produce ozone according to the following (simplified) reaction:

$$O + O_2 \rightarrow O_3 + 1.10 \text{ eV} \tag{8}$$

Ozone can also be formed in polluted atmospheres from the atomic oxygen resulting from the photochemical dissociation of NO_2:

$$NO_2 + h\nu \; (>3.1 \text{ eV}) \rightarrow NO + O \tag{9}$$

This atomic oxygen will produce ozone according to reaction (8) above.

Under natural conditions, the ozone content of the atmosphere is between 10 and 20 ppb in regions where there is no human activity. On average, this level goes up to between 175 and 300 ppb in polluted urban atmospheres and can even reach 1 ppm.

Effects

Ozone is the most toxic constituent of photochemical smogs to plants. This means that the economic threat posed by ozone is comparable to if not greater than that of *sulphur dioxide* in agricultural areas.

Its effects can be seen on numerous herbaceous and ligneous plants from concentrations of 50 ppb upwards. Symptoms include unusual pigmentation of the leaves, chlorosis or bleaching, with the final stage in the lesions being the *necrosis* of affected tissues. There are also lesions to the mesophyll cells and chloroplasts.

Even in lower concentrations than those which cause detectable damage to the aerial parts of phanerogams, ozone can decrease photosynthetic activity and slow down growth. Carnations exposed to 75 ppb for 10 days are unable to produce their inflorescences.

Furthermore, there can be synergistic effects between ozone and SO_2. Menser and Heggestad (1966) showed that 27 ppb of ozone + 240 ppb of SO_2 caused foliar lesions on the tobacco plant after only 2 hours' exposure, whereas the same plant resisted well the same period of exposure to the same concentrations of the two gases in isolation. This synergism has since been confirmed by various other research on different plant species.

Human health can also be affected by ozone. Man cannot withstand permanent doses of 2 to 3 ppm. At a level of 0.3 ppm there is irritation to the nasal and laryngeal mucous membranes, while 0.5 ppm causes a reduction in pulmonary elasticity and capacity.

6. Tobacco smoke

It would be difficult to omit from any work on ecotoxicology the problem of tobacco-smoking, a voluntarily-produced autopollution.

Composition

Tobacco smoke is a major air pollutant in western countries. It has a complex composition, containing around 3000 substances, most of which are mutagenic and/or carcinogenic. These substances are present either in a gaseous or in particle form.

The first investigations into the dangers of the different constituents of tobacco smoke concentrated on the effects of nicotine and the tars which contain many carcinogenic polycyclic hydrocarbons, among them the deadly 3,4-benzopyrene. Some of the more recent research has been directed towards the physio-toxicological consequences of inhaling carbon monoxide, which is the principal constituent of tobacco smoke, and acrolein, an aldehyde that acts as an alkylating agent and fixes on to the nucleic acids which it then changes irreversibly (Marano F., 1980).

Pathological Consequences

The epidemiological consequences of cigarette-smoking have become a real disease of modern times, especially in the urban environment where it has an additive or even synergistic action along with other air pollutants. Moreover, in contrast to the alcoholic or drug addict who only endangers his or her own health, the smoker affects the health of everyone around him or her and even of his or her children (effects on pregnant women). Mortality rates for smokers are very high in relation to the normal population. Tobacco smoking significantly reduces average life-expectancy — by about 5 years for someone who smokes 20 cigarettes a day.

The pathogenic action of tobacco smoke causes three major groups of ailments: those of the cardiovascular system, those of the respiratory system and various cancers.

(a) Effects on the cardiovascular system — The action of tobacco on this system is complex. Added to the constricting effects of the nicotine, inhalation of

the carbon monoxide in cigarette smoke causes permanent hypoxia of the myocardium. Whereas the normal level of carboxyhaemoglobin is 0.4% in the blood of non-smokers, it goes up to 5% in people who smoke 20 cigarettes a day. It has been proved that such high levels of carboxyhaemoglobin can become critical, even in the absence of any other complicating factors in patients suffering from coronary complaints.

Still more disconcerting is the fact that recent experiments suggest that smoking greatly accelerates the pathological processes of arteriosclerosis. This condition appears at an early age in smokers, particularly in the coronary arteries. In one experiment, some monkeys (*Cercopithecus*) were kept in an atmosphere containing carbon monoxide and given cholesterol in their food. They tended to have a much higher frequency of coronary stenosis than the control specimens given the same diet but kept in a pure atmosphere (Webster, 1970).

Other investigations show that laboratory animals exposed permanently to tobacco smoke develop various abnormal sterols in their blood which compare with those deposited in the arterial walls of subjects suffering from arteriosclerosis.

Under such conditions it is not surprising that the frequency of coronary disease is about 10 times higher for smokers than for the rest of the population and also that the mortality rate from myocardial infarction is three times higher in the age group of 45 to 54 years for people who smoke at least one packet of cigarettes a day.

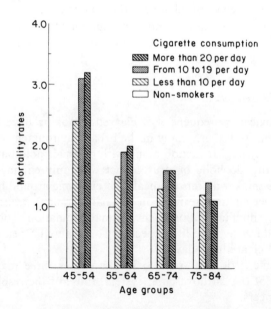

Fig. 3.38 Mortality from myocardial infarction according to the numbers of cigarettes smoked and the age group. (*In* Masters, 1971)

(b) Smoking and chronic bronchitis—The role of smoking as a cause of chronic bronchitis is notorious. This condition is eight times more likely to occur in adult smokers, whatever the age group, than in non-smokers. There is also a clear additive effect between smoking and SO_2 air pollution in the genesis of chronic bronchitis (cf. Fig. 3.39).

(c) Carcinogenic effects—Tobacco smoke plays a significant part in inducing several types of cancer. Apart from its influences on the development of bronchopulmonary carcinomas, it is known to have a positive correlation with the formation of certain gastric, hepatic, renal and vesical cancers.

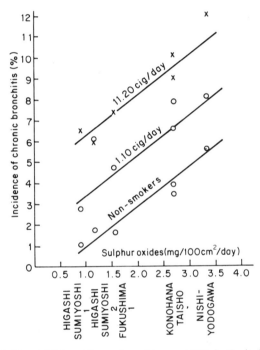

Fig. 3.39 Additive effects of smoking and SO_2 in the incidence of chronic bronchitis among inhabitants of various Japanese town and cities. (After Nishiwaki *et al.*, 1970. *Reproduced by permission of the International Atomic Energy Agency, Vienna*)

A recent Polish statistic suggests that out of 1000 cases of cancer of the larynx, 990 were smokers. Bronchopulmonary cancers were rare at the beginning of this century but have increased alarmingly ever since. Although atmospheric pollution has an undeniable effect on their induction, smoking is still the most important cause (17 out of 20 cases at the present time).

The epidemiological study of the role of tobacco smoke in the genesis of bronchopulmonary cancers has been complicated by the long latency period, sometimes of about 20 years, which generally occurs between the first inhalation of the smoke and the appearance of clinical symptoms.

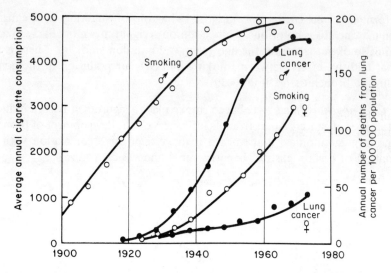

Fig. 3.40 Correlation between the growth in smoking and increase in mortality from lung cancer in Great Britain since the beginning of this century. These graphs also show a latency period of about 20 years (at least) between the start of exposure to the carcinogens in tobacco smoke and the appearance of the disease. (After Cairns, 1975. Reproduced by permission of Scientific American, Inc.)

Studies carried out in Great Britain give an impressive demonstration of this time lapse. They also show that, contrary to popular opinion, men are no more susceptible than women to this disease. This is because women took up the smoking habit at the end of the Second World War, whereas men started mainly at the beginning of this century. Consequently, in the British Isles the increase in frequency of bronchopulmonary cancers in men started between about 1925 and 1935 and in women around the middle of the 1960s (Fig. 3.40).

The harmful action of the carcinogenic agents contained in tobacco smoke is enhanced by nicotine, which inhibits their metabolism. Weber *et al.* (1974) observed that a single dose of nicotine given to rats causes biliary excretion of the 3,4-benzopyrene to decrease, linked to inhibition of the benzopyrene hydroxylase of the liver, lung and small intestine.

(d) Effects on the foetus — A foetus can become the victim of smoking habits if its mother smokes. The amount of oxygen carried to the baby is decreased due to the high levels of carboxyhaemoglobin in the mother's blood.

The average weight of newborn babies whose mothers smoked during pregnancy is less than normal and the mental development of these children is lower by a few months at the age of 7.

In conclusion, it is astounding how ineffectual or perhaps unwilling our public authorities seem to be to put an end to this voluntary form of pollution. Their laxity is even less justifiable if one considers the human suffering, large number of early deaths and enormous social costs caused by this disease, which is less spectacular but just as destructive as alcoholism. If stringent measures can be taken to control narcotics how long will it be before the cultivation of tobacco in so-called 'developed' countries is forcibly banned?

7. Lead

Of all the heavy metals contaminating the biosphere, lead is, after mercury, the most important so far as its ecotoxicological effects are concerned.

Although biogeochemical phenomena naturally carry 180 000 tonnes of lead into the oceans annually, man was already extracting some 2×10^6 t/year of lead from the lithosphere during the 1970s.

Sources

The main cause of lead pollution at present is its use in the form of alkyl lead in petrol as an anti-knock agent. The combustion of motor-fuels containing tetraethyl lead does in fact release particles of lead into the atmosphere. Every vehicle emits an average of 1 kg/year of lead into the air in the form of inframicroscopic particles.

Since 1923, the date when alkyl lead was first added to petrol, pollution of the ecosphere by lead has increased considerably: by 1970 the world-wide

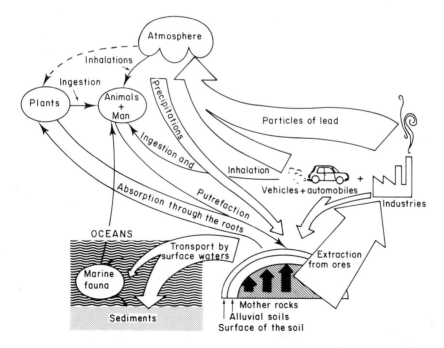

Fig. 3.41 The biogeochemical cycle of lead. (After Jenkins, 1975. *Reproduced by permission of the Société Nationale de Protection de la Nature*)

production of these compounds was up to some 350 000 tonnes per year (in lead equivalent). A study made by Murozumi *et al.* (1969) of the lead content of the annual layers in core samples of snow in Greenland showed that the lead content had gone up suddenly more than twentyfold over the past 50 years. Quite logically, the scientists explained their observation by relating it to the start of tetraethyl lead manufacture from 1923 onwards.

The scale and extent of contamination of the biosphere by lead becomes evident when its content is measured in biocoenoses situated in remote parts of the Earth.

The Contamination of Ecosystems by Lead

Research carried out by Hsiao and Patterson (1974) on one of the most isolated ecosystems in the United States, in the valley of Thompson Canyon in the High Sierra of California, provides a vivid illustration of this phenomenon. They proved that most of the lead found in the voles (*Microtus montanus*) and the sedge (*Carex*) on which they feed is entirely exogenous and comes from the road traffic in the polluted areas of southern California several hundred kilometres away.

In food chains which have not been interrupted by man, various mechanisms prevent lead from following all the other biogenic metals further on up the chain towards the higher trophic levels. At each of these levels, the elimination of lead can be evaluated quantitively by the ratio of Pb/Ca. In the food chain studied by Hsiao and Patterson, this ratio went down by 200 times between the rocks making up the substrata of the soils for the *Carex* and on reaching the voles. Normally, the ratio would have gone down by 1200 times if polluted lead aerosols had not been deposited on the leaves of the *Carex*, which then increased the lead intake of the voles. (Table 3.13)

Various experimental data prove the technological origin of the lead deposited on the aerial parts of the plants on which the voles feed. For instance, the ratio

Table 3.13 Molecules of metals per 10^6 calcium molecules contained in a trophic chain of Thompson Canyon (California). (After Hsiao and Patterson, 1974, *Science*, **184**, 989–994. *Copyright 1974 by the AAAS*)

Metal	Rocks	Interstitial water in the soils	*Carex scopulorum*	Voles
Calcium	1 000 000	1 000 000	1 000 000	1 000 000
Baryum	15 000	3 800	2 000	330
Lead	280	210	54	1.4
After washing the leaves in acid to destroy the deposits				
Lead	280	210	9	0.2

Pb/Ca + K shows that about 97% of the lead particle deposits in the Thompson Canyon are from industrial sources. The ratio is 2 in the Los Angeles air, it is 0.01 in the foliar deposits in the Canyon, and only 0.0005 in the Sierra rocks. These conclusions are confirmed by a study of the isotopic ratio of ^{206}Pb/^{207}Pb in the urban atmosphere and in the rocks and foliar deposits in the Canyon.

At the present time, environmental pollution by lead has not yet reached a high enough level to endanger terrestrial biocoenoses. Only man and anthropophilic animals are threatened, mainly in urban environments, by the atmospheric pollution there and because of certain technological uses of lead which cause it to contaminate some foodstuffs.

In the United States, the average level of urban air pollution by lead is between 0.1 and 3.4 μg/m^3. In the Soviet Union, the maximum level allowed in cities is 0.7 μg/m^3. As a general rule, industrial legislation in European countries permits 200 μg/m^3 of lead in the workplace for a 40-hour week.

Pigeons living in cities, spending much of their time at ground level, contain very high levels of lead in their bodies. In Paris, the rock-pigeons (*Columbia livia*), with their hidden lead-poisoning have become good indicators for measuring atmospheric pollution (Jenkins, 1975).

An important source of lead contamination of the human body is foodstuffs. Tobacco smoke also contains quite significant concentrations of the metal (Table 3.14). Table 3.14 shows that in the United States the daily intake of lead is 40% higher for a non-smoking city-dweller than for people living in rural areas.

Some analyses done in Manchester, Great Britain, show that the average lead content of the serum of children goes up to 0.31 ppm and even goes above 0.8 ppm in 4% of those sampled, which is a concentration where cerebral lesions can occur. Lead does in fact accumulate in the brain and can cause serious encephalopathies. In addition, permanent exposure of humans to weak doses of lead in the long term causes anaemia, renal malfunctions, and various endocrine disorders, especially of the reproductive glands.

Table 3.14 Average absorption of lead by a 'normal' individual in the United States. (*In* Simmons, 1974. *Reproduced by permission of Academic Press*)

Source of contamination	Daily intake	Concentration of lead in the source	Lead ingested (in mg/day)	Fraction absorbed	Lead absorbed per day (in mg)
Food	2 kg	0.17 ppm	330	0.05	17
Drinking water	1 kg	0.01 ppm	10	0.1	1
Air { urban	20 m^3	1.3 mg/m^3	26	0.4	10.4
Air { rural	20 m^3	0.05 mg/m^3	1	0.4	0.4
Cigarette smoke	30 cig/day	0.8 mg/cig	24	0.4	9.6

II — Other Atmospheric Pollutants

1. Dusts

There are two groups of dusts according to their size:
—large particles, above 0.1 μm in size;
—small particles, wrongly called aerosols.

Origin and Composition

'Aerosols' are particles less than 0.1 μm in size. The smallest of them, Aitken nuclei, measure just 300 Å in diameter on average and can be as tiny as 10 Å. Although a certain number of particles, such as smoke and lead 'aerosols' are products of combustion, the majority are from other sources. The iron and steel industry, mining industries, cement works and large-scale civil engineering projects are the principal sources of dust. Even in the urban environment, only 5% of the dusts come from combustion.

Dusts collected in industrial zones contain numerous varieties of minerals, mainly: quartz, calcite, feldspar, gypsum, anhydrite and asbestos. The last substance on the list is a hydrated magnesium silicate with many different uses from the manufacture of brake linings to fireproofing material in metallurgy or as insulation material in the construction industry, etc.

The dusts also contain a large number of non-volatile metals and metalloids either in a pure state or in the form of salts or oxides.

In the Chicago region, Brar *et al.* (1970) found 20 elements in detectable quantities among the aerosols. In descending order, these were: iron, aluminium, zinc, manganese, sodium, chromium, vanadium, molybdenum, arsenic, antimony, etc.

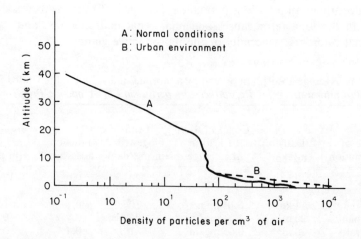

Fig. 3.42   The distribution of particles in the atmosphere according to altitude. (After Varney and MacCormac, 1971. *Reproduced by permission of D. Reidel Publishing Co.*)

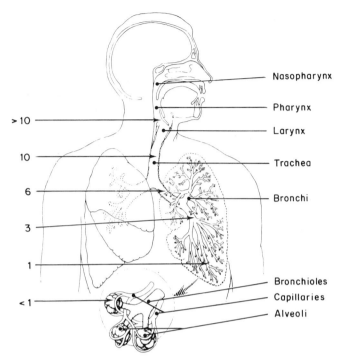

Fig. 3.43 Contamination of the human respiratory system according to the size of particles inhaled (the particle sizes are shown as diameters expressed in μm). (*In* Masters, 1971)

The last category of particles from a technological origin are the fluorides. These are emitted principally by alumina electrochemical industries, superphosphate factories and brickworks.

The Effects of Particles on Living Beings

In some heavily industrialized regions about 300 t/km²/year of dusts can be deposited (*in* Mellanby, 1971).

Particles falling on plant leaves can have phytotoxic effects. They not only reduce photosynthetic activity but also impede the germination of pollen by covering the flowers' stigmata. The dust from cement works, being highly alkaline, causes foliar chlorosis.

Human health is greatly affected by air pollution from dusts. Fortunately, the largest particles are filtered out in the rhinopharynx and trachea, but particles of less than 6 μm in diameter can reach the bronchi, and those smaller than 1 μm can enter the alveoli. A whole series of medical conditions can result, some of which are extremely serious.

(a) Allergies—Numerous allergies can be produced by aeroallergens. Most of them involve hygroscopic solid particles with a diameter of between

1 and 80 μm. When such dusts reach the alveolar epithelium they cause asthma.

(b) Chronic bronchitis—This results from continuous exposure to various gaseous or solid air pollutants. Symptoms are a dry cough and bronchial hypersecretion, followed by progressive pulmonary hypoventilation and, in the long term, by heart failure. Although both tobacco smoke and sulphur dioxide are other important causes of chronic bronchitis, it has been proved that the illness occurs more frequently and more seriously in regions where the atmosphere contains over 100 $\mu m/m^3$ of particles. Pulmonary emphysema, characterized by dilation and loss of elasticity of the alveoli, is often associated with chronic bronchitis.

(c) Lung cancer—The contribution made by dusts to the induction of lung cancer in non-smoking city-dwellers is explained by the presence of carcinogenic polycyclic hydrocarbons and other substances with similar effects in the dusts.

Truhaut (1960) demonstrated the carcinogenic potential of samples of dust taken from the Paris air. He was able experimentally to cause skin cancers in mice by painting them with the dust. Hickey (1971) stresses the mutagenic properties of various gaseous or solid atmospheric pollutants.

(d) Silicosis—Silicosis or pulmonary fibrosis results from inhaling silica and silicates. It is caused by histiocytes which accumulate in the pulmonary parenchyma and store the mineral particles in their cytoplasm which are transformed into fibrocytes. This leads to hardening of the alveoli and loss of their elasticity.

(e) Asbestosis—This term covers a range of conditions of varying severity all linked to the inhalation, or very rarely, the ingestion of asbestos dusts. Minerals of the asbestos group are magnesium silicates (chrysotile, for example) or silicates (crocidolite) with a fibrous structure. Asbestos material is produced by mechanical dissociation of the bundles of mineral fibres found in natural ore.

At the present time, asbestos has a large number of industrial uses. It is employed in building and construction, partition wall manufacture, roof coverings in fibrocement, as a thermal insulator for both domestic and industrial buildings, and as fireproofing material, etc. It is also used in engineering, for instance in the manufacture of clutches and brake-linings. The amount of asbestos mined world-wide to cater for all these needs is some 4 million tonnes every year.

Although the first cases of lung cancer caused by asbestosis were recognized back in the 1930s in asbestos miners, it was not until 1968 that Selikoff and his colleagues were able to prove formally the role of asbestos in inducing lung cancer. They confirmed that asbestos miners who were also smokers were 60 times more likely to die of lung cancer than people who were non-smokers and not exposed to asbestos (Table 3.15). This study also pointed out the

Table 3.15 Mortality from lung cancer in asbestos miners. (After Selikoff et al., 1968.)

Smoking habits	Number of individuals	Number of deaths observed	Theoretical probability of death from lung cancer, calculated in relation to the whole population
Individuals who had never smoked	48	0	0.05
Pipe or cigar smokers	39	0	0.13
Cigarette smokers	283	24	3.16

importance of synergistic effects between the two carcinogens. It further established a very clear positive correlation between the number of incidences of the disease and the average asbestos dust content of the air to which the subjects were exposed before ever showing clinical symptoms of cancer.

Asbestos, acting as a carcinogen, can induce several types of cancer, but the most frequent of these, resulting directly from exposure to the material, is a very distinctive type called a pleural mesothelioma. It involves a diffuse fibrous infiltration of the alveolar parenchyma and pleura, taking the form of tumours.

A study was made of patients suffering from mesothelioma who had lived near an asbestos factory but never worked in the industry. It showed that there is an extremely long latency period for the disease, of an average 37 years. In extreme cases, 50 years could pass between initial exposure to this carcinogen and appearance of the mesothelioma.

Asbestos contamination of the urban environment is now of great concern in certain places, due to its extensive use in the building industry over the last decades (Newhouse, 1977, in Lenihan and Fletcher, 1977). Quite recently, chrysotile fibres were found in food and drink that had been filtered through asbestos filters; this event caused a considerable uproar in developed countries.

However, even though it has been proved that there is a greater incidence of gastrointestinal tumours in asbestos miners and other high-risk categories of workers, there is no clinical indication as yet of any pathological risk to the consumer. Nevertheless, it is essential to reduce drastically any use of asbestos which might entail environmental contamination from inhalation or ingestion, in view of the available data on the dangers of this substance.

2. Fluorine

Freons

Atmospheric pollution by fluorine comes from various technological sources. Apart from the emission of fluoride particles from certain industries, as described earlier, the release of freons is the other important contributor to this contamination. Freons are extremely volatile chlorofluorine hydrocarbons with multiple uses, for example as coolants in refrigerators and air conditioners or as propellants in common household aerosol sprays such as deodorant and hairsprays. World production of freons was close to 1 million tonnes in 1980. Now they are suspected of attacking the stratospheric ozone layer and cause global concerns (cf. for instance, Dupas, 1975 and National Academy of Sciences, 1979). However, for the moment they are not of primary ecotoxicological concern as they are only negligibly toxic, except in very high concentrations when they can cause heart trouble.

Fluorides

Contamination of forests and agricultural ecosystems by fluoride particles is, however, of grave concern at the present time. When they are deposited on the soil surface, mineral fluoride compounds infiltrate the trophic chains and can cause serious disorders in both plant and animal communities.

Effects on Plants

The gaseous fluorides (FH, SiF_4, H_2SiF_6) or solid fluorides ($F_6Al\ Na_3$, F_3Al, F_2Ca, FNa, fluorapatite) penetrate plants either through the stomata or directly through the leaves. They can also be absorbed through the roots, particularly in heavily fluoride-polluted soils. As fluorine plays a marginal physiological role in plants and cannot be metabolized, it can accumulate in large amounts especially in the foliar system. When the fluoride concentration reaches a certain level, varying with different plant species, foliar lesions develop. These are sometimes characterized by a chlorotic appearance, and in other cases by a greenish-grey colouring of the parenchyma before the necrosis develops. The whole of the foliar lamina takes on a brownish colour with a darker line marking the edge of the affected area from the rest of the leaf. In conifers, the necrosis has a browny-red colour and spreads towards the base of the needle.

There is a great variation in the toxicity of fluorine to plants. The most sensitive plants can be affected after a week's exposure to a concentration of 0.4 to 1 μg of F/m^3, (1 ppb of fluorine = 0.8 μg of F/m^3 of air), which is about 1 ppb of this element (MacCune, 1969, *in* Linzon, 1971). The most tolerant plants develop necrosis only at concentrations twenty times greater than these.

Among the most sensitive species to fluorine are the Liliaceae, the gentians, the arborescent Rosaceae (*Prunus, Amygdalus* in particular), the conifers and the vine. Whereas gladioli (*gladiolus* sp.) develop necrosis once they have 20 ppm of fluorine in their leaves, there are certain hickory trees (*Carya* sp.) that can contain up to 1000 ppm in their foliar lamina without any apparent damage.

Under the microscope, affected parenchymatous cells show granulations, vacuolization and eventually total plasmolysis. In conifers, the resin canals become blocked due to hypertrophy of their parietal cells.

Fluorides also have a significant inhibiting effect on the photosynthetic activity of plants in much smaller concentrations than those causing foliar lesions. They can also inhibit enolase, an enzyme essential for glycolysis in plants.

The damage caused by fluorine pollution to forests can be very extensive. In France, several thousand hectares of conifers have been destroyed in the Maurienne valley (Alps) where, since 1960, some 1200 hectares of *Pinus sylvestris* have disappeared from an area round an aluminium electrochemical plant.

Effects on Animals

Domestic animals given feed containing fluorine can suffer long-term poisoning called fluorosis. In fact fluorine, apart from its intrinsic cytotoxic properties and due to its affinity for calcium, disrupts the ossification process. It is, therefore, one of the substances whose dose–response curve shows an area favourable to the organism and a toxic area (cf. Fig. 1.10). Fluorine is necessary in low doses for ossification in vertebrates, and when it is ingested in trace amounts with food it increases the strength of bone tissue and resistance to dental caries. However, if taken in excessive doses it can cause fluorosis, a disorder with various symptoms of increasing severity. The first symptoms are dental: the teeth become weaker and mottled. Then bone deformities develop, followed by progressive cachexia which can be fatal to the affected animals.

The maximum fluorine concentration that can be tolerated by cattle in their feed is between 30 and 50 ppm, and this concentration can go up to 100 ppm for sheep and pigs and to 300 ppm for domestic poultry (*in* Lillie, 1970). In cows, poisoning from minute doses causes a reduction in milk production accompanied by a lowering of the number of lipids in the milk.

Another ecological consequence of fluorine pollution is the damaging effect it can have on the entomofauna. Fluorine is actually very toxic to most orders of insects. Bees are particularly sensitive to it and no apiary can survive in polluted zones.

III POLLUTANTS OF THE HYDROSPHERE

A certain number of mineral, organic or fermentable substances are contaminants exclusive to limnic ecosystems and to the oceans and only circulate in the direction from continent to hydrosphere. Among these, the most important are the hydrocarbons and detergents. Others include numerous solid wastes dumped by various industries into continental or marine waters: 'deads' from mines, red and yellow sludges, etc., some of which have caused concern recently. Another type of pollution which essentially affects the limnic environment is the discharge into the superficial waters of fermentable organic matter and nutrient salts.

The resulting phenomenon of dystrophication (mistakenly called eurtrophication in certain publications) is menacing most of the lakes in industrialized countries.

I—Marine Pollution

There exists a whole series of polluting agents exclusive to the marine environment or which, if they are also found in continental waters, exercise most of their damaging effects on oceanic organisms.

1. Pollution of the oceans by hydrocarbons

Causes

The voluntary or accidental discharge of oil into the sea is a primary cause of contamination of the hydrosphere on a global scale. Not only does this pollution have worrying consequences, especially in areas of the continental shelf where the most productive fisheries are located, but it is also threatening the ecological balance of land-locked seas, even large ones like the Mediterranean.

The fact that there are marine zones where oil tankers can legally discharge their tanks after cleaning constitutes a real challenge to ecological teaching. It is estimated that 0.5% of the petroleum transported by tankers is discharged more or less legally into the sea during tank cleaning. In reality, the zones where unballasting is prohibited are not properly respected by the tankers, especially by ships sailing under flags of convenience. Because of this, the Channel and the western Mediterranean are continually being contaminated by the illicit unballasting of these tankers.

In total, the world's oceans were receiving an annual tonnage of hydrocarbons estimated at 6.1×10^6 tonnes at the beginning of the last decade, according to the National Academy of Sciences, US (1975). At the present time, the extension of the load on top procedure to more than 80% of the world's tanker fleet and various recuperation measures taken following the increase in crude oil prices have reduced the amount of petroleum dumped. At the beginning of the 1980s, from 4×10^6 tonnes to 4.5×10^6 tonnes of hydrocarbons per year were being discharged into the oceans, including those from natural causes (seepages from offshore deposits). As one tonne of crude oil covers some 12 km^2 of ocean surface as a very thin, almost molecular film, the whole of the world's oceans are now permanently contaminated by petroleum.

At least 50 oil tankers per year were involved in navigation accidents during the 70s. The disasters of the *Olympic Bravery* (March 1976), of the *Urquiola* (May 1976) and of the *Amoco Cadiz* (March 1978) were sad illustrations of the dramatic consequences of such accidents. In addition, faced with the exhaustion of continental oil supplies, the increase in the number of offshore wells will only add to the risk of leaks and attendant pollution. For instance, the catastrophe of the Ixtoc One well in 1979 resulted in nearly 1 million tonnes of petroleum escaping into the Gulf of Mexico, as the situation took more than 6 months to control.

The Nature and Forms of Marine Pollution by Hydrocarbons

At the present time, most of the surface of the world's oceans is polluted by

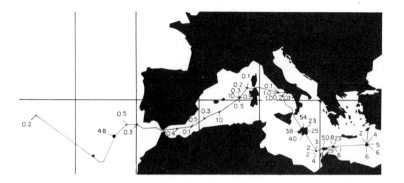

Fig. 3.44 Evaluation of the extent of contamination by oil slicks of the Mediterranean and the western Atlantic. Each point is of a diameter proportional to the volume of 'tar balls' found at each sampling. It represents the product of 100 times this volume expressed in cubic centimetres. In reality, none of the samples taken during this oceanographic survey between Lebanon and the Azores was totally free of petroleum residue. (After Horn et al., 1970, Science, **168**, 245–246. *Copyright 1970 by the AAAS*)

the different chemical types of hydrocarbons contained in crude oil and by the by-products from their biogeochemical degradation.

There are important variations in the composition and physico-chemical properties of crude oils according to their origin, although they all have the same groups of basic constituents.

Hydrocarbons form the largest chemical group in the composition of petroleum. They can be subdivided into unsaturated aliphatic hydrocarbons, called olefins, which are rarely found in crude oil but appear after catalytic cracking, and saturated aliphatic hydrocarbons with a straight or branched chain, which make up the biggest group. Petroleum also contains aromatic and heterocyclic hydrocarbons. The aromatics include benzene and the corresponding polycyclic compounds. The heterocyclic hydrocarbons, also called naphthenic hydrocarbons, are usually saturated compounds containing complex, non-benzene, cycles. Also found in crude oils are sulphur compounds (mercaptans, for example), hydroxyls (phenols) and nitrates. Though they are not present in large amounts, they play an important part in the toxicity of these substances. The more or less polymerized part of this group contains the asphaltenes which, along with long-chain aliphatic compounds (paraffins) and certain naphthenic hydrocarbons, constitute the heavy fractions of crude oil.

When petroleum is discharged into the sea, various physico-chemical processes intervene to disperse it: surface-activity, evaporation, emulsion and dissolving all combine their effects to spread the oil slick into a thin film over the surface of the sea, covering huge areas (sometimes thousands of square kilometres in the case of major disasters). The formation of aerosols by wind action can carry the hydrocarbons to the coast and pollute the fields, as happened at the time the *Amoco Cadiz* was wrecked. Evaporation eliminates the volatile and most toxic fractions. In contrast, photo-oxidation can transform certain hydrocarbons

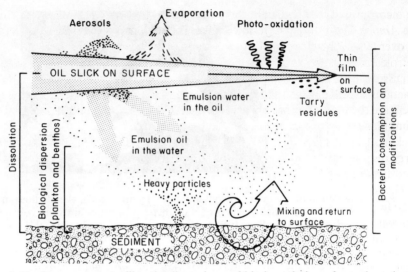

Fig. 3.45 Processes controlling the dispersion and biodegradation of petroleum in the marine environment. (After Clark and MacLeod, 1977. *Reproduced by permission of Academic Press*)

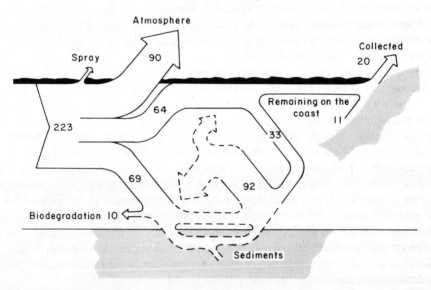

Fig. 3.46 What became of the petroleum after the *Amoco Cadiz* disaster? Figures are expressed in 10^3 tons (After Marchand *et al.*, 1979)

into aldehydes and other compounds which are far more dangerous than the original products.

The surface action of the sea contributes to the formation of emulsions of oil in the water and water in the oil (cf. Fig. 3.45). One part of these emulsions, whose formation is aided by the dispersants used to break up oil slicks, will be incorporated into the sediments, along with the heavy particles. Following the *Amoco Cadiz* disaster, petroleum was found in benthic sediments at depths of over 80 m. The movement of the waters in the ocean depths can lead to the petroleum returning to the surface after weeks or even months.

Dissolving is another important process, as virtually all of the petroleum fractions dissolve in sea-water, though with varying rates of solubility. Then there are the biological factors helping the dispersion of the petroleum, ranging from planktonic organisms to benthic animals and nekton. Finally, after the slick has been dispersed, only a thin superficial film of hydrocarbons and some tarry residues will remain which will be subject to bacterial consumption and modification.

The residual products of decomposition bind together after a few weeks and become tarry nodules or lumps of irregular shapes measuring between 0.1 and 10 cm in diameter which float on the ocean surface before being washed ashore, a well-known problem for bathers.

It has been possible to measure the extent of the degradation of hydrocarbons by the aerobic bacteria which cover the lumps in a thin greyish layer. The oxygen consumed by this biodegradation has been estimated at 12.5 mm^3/hour/cm^3 of tar by Horn *et al.* (1970). A few particularly tolerant species of marine invertebrates use these tar lumps as substrate. The pelagic isopod *Idotea metallica* is often found fixed on these lumps, as are the cirripeds *Lepas anatifera* and *L. fascicularis*.

The frequency of these nodules even in the most remote pelagic zones shows the present extent of contamination of the oceans by petroleum. During an oceanographic survey between the island of Rhodes and the Azores, Horn *et al.* found such tar lumps in 75% of the samples taken with a neuston net. It was possible to calculate that there is an average number of 1 mg/m^2 of tar lumps in the Atlantic and 20 mg/m^2 in the Mediterranean (Morris, 1971).

Ecological Consequences of Petroleum Pollution of the Oceans

There is now a considerable amount of experimental data available on the ecotoxicological effects of petroleum. The results have either been obtained in the laboratory on isolated test species or artificially reconstituted communities, or from synecological studies carried out following the wrecking of oil-tankers that have caused wide-scale marine pollution. The disasters of the *Torrey Canyon* and, more recently, the *Amoco Cadiz* have enabled scientists to gather precious information on the biocoenotic consequences of marine pollution from petroleum products (Smith, 1968).

(a) Effects on the phytoplankton and algae—Petroleum pollution has a disastrous effect on most species of phytoplankton and benthic algae.

After the *Torrey Canyon* disaster, the microplankton from the Prasinophyceae group (*Halosphera* and *Pterosperma*) living in the superficial waters of the ocean were found to have been particularly affected by the hydrocarbons. Most

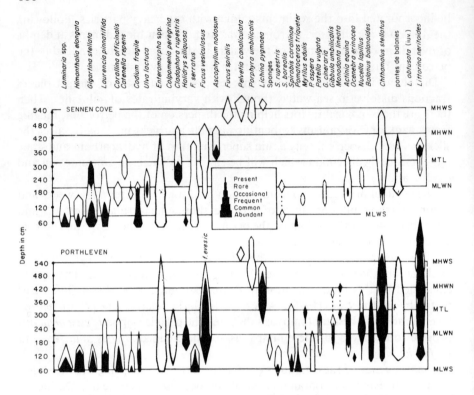

Fig. 3.47 Effects of the petroleum spilt by the *Torrey Canyon* on the biocoenoses of the intertidal zone in Cornwall. The species are shown as a function of depth as follows: in grey are the species which developed following the oil slick; in white are the species which existed before the disaster, with the distribution of these same species shown in black six months after the disaster. (After Nelson-Smith, 1970. *Reproduced by permission of Cambridge University Press*)

of the species of algae growing in the intertidal zone off the Cornish coast also suffered and their numbers decreased significantly (cf. Fig. 3.47). However, a rapid repopulation was also observed in certain species, especially the *Fucus*. A similar phenomenon was also observed in California where some *Macrocystis* increased their volume dramatically in the months following accidental pollution of the coastline by hydrocarbons. After the *Torrey Canyon* some algae, such as *Colpomenia peregrina* and some *Enteromorpha* sp., not previously found in the contaminated areas, were also found to be multiplying rapidly.

It has been demonstrated experimentally that when extremely thin films of crude oil, of less than 10 μm in thickness, cover over surface seaweeds (*Fucus vesiculosus, Laminaria digitata, Porphyra umbilicalis*), they reduce the diffusion of gases and inhibit the algae's photosynthesis.

In addition, petroleum can decrease or even inhibit the reproductive capacity of the seaweeds, even in very weak doses. For instance, 0.2 μg/l of heavy fuel

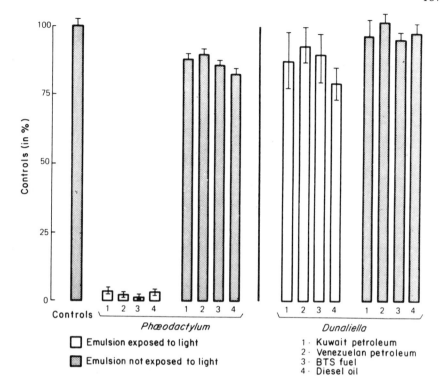

Fig. 3.48 The influence of phenomena of photo-oxidation on the toxicity of emulsions of petroleum dispersants to phytoplanktonic organisms. These histograms show the level of primary production, expressed in percentages, as against the non-treated control specimen. It is found that at a dose of 50 mg/l of dispersant the primary production of *Phaeodactylum* is almost completely inhibited. (Lacaze, 1978. *Reproduced by permission of Pergamon Journals Ltd.*)

oil No. 2 or crude oil is sufficient to inhibit the germination of zygotes of the Fucaceae *Fucus edentatus* (after Steele, 1977 *in* Lacaze, 1980).

The primary production of phytoplankton is definitely reduced by petroleum pollution of the oceans and its associated phytotoxic effects, as shown in a study by Lacaze (1978). Various experiments on microplanktonic communities, on phytoflagellates (*Phaeodactylum, Dunaliella*) and marine diatoms (*Amphora, Navicula*) showed that the photosynthetic fixation of ^{14}C was significantly altered by petroleum concentrations of less than 1 ppm. Also, the photo-oxidation of emulsions of petroleum in water, obtained by the adding of a dispersant, produces substances that have an even more drastic effect on primary production (cf. Fig. 3.48).

The different hydrocarbons making up crude oil do not have the same effects on the photosynthetic activity of phytoplankton. The study by Lacaze (1978) on the influence of four different aromatic compounds on *Phaeodactylum tricornutum* shows that naphthalene has a greater inhibiting effect than the other hydrocarbons from this group (cf. Fig. 3.49).

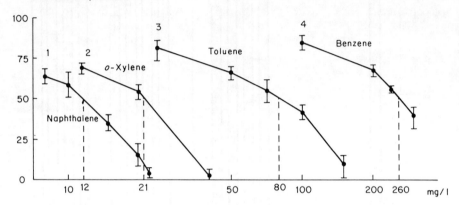

Fig. 3.49 The influence of four aromatic compounds on the photosynthetic activity of the microalga *Phaeodactylum tricornutum*. (*In* Lacaze, 1980. Reproduced by permission of Pergamon Journals Ltd.)

The same author provided more evidence of petroleum affecting primary production in a study of pollution by Kuwait crude oil of an estuarine zone. Although production was some 20 mg $C/m^3/h$ in the control ecosystem, it fell to 10 mg $C/m^3/h$ in the contaminated zone after 24 hours and was reduced to virtually nil between the 5th and 15th days following exposure to the pollution. One month after the beginning of the experiment, the productivity of the contaminated zone was still significantly weaker than the control area (Lacaze, 1974).

(b) Effects on the zooplankton—Petroleum pollution also has serious effects on zooplankton. After the *Torrey Canyon* disaster they were found to have disappeared completely from the polluted areas. In addition, some 90% of the pelagic eggs of pilchards were destroyed in the most heavily contaminated zones. The fry also died in large numbers, with no more than 5 per 100 m^3 of water surviving whereas in non-polluted waters fry are usually present in quantities greater than one per square metre.

(c) Effects on marine animals—Various research studies on different organisms have made it possible to measure the toxicity of the different hydrocarbons contained in crude oil for marine metazoa. Table 3.16 gives examples of the toxicity of some aromatic derivatives which are particularly poisonous to the sea bass and some marine invertebrates.

Even if fish and other marine animals are not killed by the hydrocarbons, petroleum contamination can have serious ecotoxicological repercussions as it involves food chains eventually leading to man. One example was that of the *Scomberesox saurus* or needlefish, an abundant species in temperate waters and of great importance to the canning industry, which was found to have swallowed tarry lumps in its search for food.

Mussels gathered on coasts near to industrial regions contain between 5 and 10 times more hydrocarbons than those from non-polluted zones. The 'taste of petroleum', even if it does not necessarily make fish toxic, can prevent its

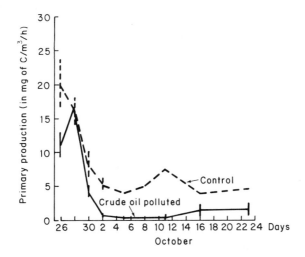

Fig. 3.50 The influence of petroleum pollution on marine primary production. Carbon fixation (in mg/m^3/hour) was measured in two neighbouring estuarine zones, one a control area and the other deliberately contaminated by Kuwait crude oil for a period of a month. The vertical bars show the length of the confidence interval for 95% reliability. (After Lacaze, 1974)

Table 3.16 The comparative toxicity of various aromatic hydrocarbons to certain marine metazoa (LC$_{50}$ after 96 hours expressed in mg/l). (After Rice et al., 1977)

HYDROCARBONS	LC$_{50}$ after 96 hours (in mg per litre)				
	Polychaeta (1)	Shrimp (2)	Crab (3)	Shrimp (4)	Sea bass (5)
Benzene	—	27	108	20	7.8
Toluene	—	7.5	28	4.3	7.3
Ethyl benzene	—	—	13	0.5	4.3
Ortho-xylene	—	—	6	1.3	11
Para-xylene	—	—	—	2.0	2.0
Dimethyl naphthalene	2.6	0.7	0.6	—	—
Methyl phenanthrene	0.3	—	—	—	—

(1) *Neanthes arenaceodenta,*
(2) *Palaemonetes pugio*
(3) *Cancer magister,*
(4) *Crangon franciscorum*
(5) *Morone saxatilis*

commercial sale and lead, therefore, to economic loss for fishing-grounds affected by pollution.

The consequences of petroleum pollution for the fauna from intertidal zones and benthic environments were studied in detail following the *Torrey Canyon* disaster in 1967 (cf. for example Smith, 1968 and Nelson-Smith, 1970) and later after the *Amoco Cadiz* was wrecked in 1978 (cf. Hess, 1978 and Marchand *et al.*, 1979).

The *Amoco Cadiz* was transporting, from the Persian Gulf, 223 000 tonnes of light petroleums containing weak amounts of asphalt but particularly rich in aromatic compounds. These accounted for between 30 and 35% of the total and were both extremely toxic and virtually non-biodegradable.

After the tanker spilled its cargo into the sea near Portsall, more than 150 km of the Brittany coast was polluted to varying degrees, but all serious. Not only were the intertidal zones severely contaminated by the crude oil or some of its constituents, but also the benthic environments. Two months after the catastrophe, hydrocarbon concentrations of more than 30 ppm were found in sediments 32 km from the shore and at depths of over 80 m (cf. Fig. 3.51).

In the Bay of Lannion alone, over an area of submerged coastline equivalent to 10 km^2, some 3.5 million dead Solenidae were counted, along with 10 million dead *Echinocardium cordatum* (Echinidae), 7.5 million *Cardium edule* and other bivalves, of which 5 million were *Mactra corallina* and *Donax vittatus*.

However, the macrophytic algae (*Laminaria, Fucus*, etc.) and the fauna from the rocky facies in the intertidal zone were less affected proportionally than they had been at the time of the *Torrey Canyon* disaster, due to the more restricted use of detergents.

A study of the evolution of the intertidal fauna in the two years after the *Amoco Cadiz* showed that the initial period of massive mortality was followed by a repopulation phase varying in success from one species to another (Raffin *et al.*, 1981). It appeared that certain dominant species in the rocky facies, such as the *Patella* (*Patella vulgata*), now had a disrupted demoecological structure, with a smaller number of older organisms and a reduced density compared with the initial phase (cf. Fig. 3.52).

(d) Effects on birds—Sea-birds are greatly affected by hydrocarbon contamination of the oceans. The colonies of Laridae (gulls) and Alcidae (auks) on Cap Sizun were decimated by the oil-slick from the *Olympic Bravery*. The Procellariiformes (petrels), auks, gulls and Anatidae (swans, geese and ducks) often land on the patches of oil and their feathers become irreversibly clogged with oil. The hydrocarbons have a great affinity for the waterproofing oils in the birds' feathers and they destroy this protective layer, so essential for keeping the bird warm. Birds covered in crude oil die of congestion as they are no longer protected from the cold. In addition, they are poisoned by the petroleum they ingest when diving or trying to clean their feathers. This causes, among other effects, severe perturbations of the endocrine system, particularly of the adrenal glands.

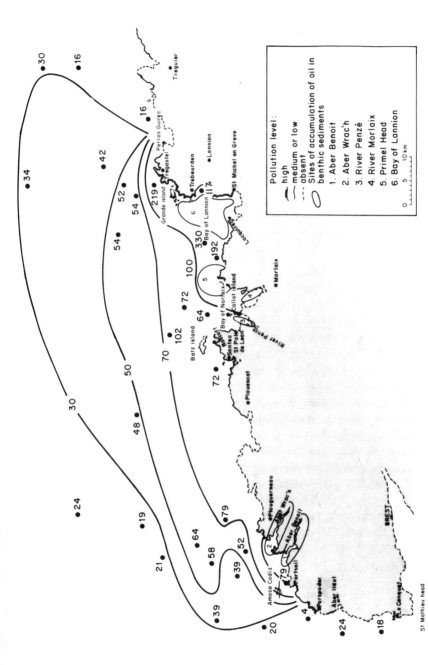

Fig. 3.51 Pollution of the Brittany coast from the *Amoco Cadiz* catastrophe in March 1978. The figures represent the hydrocarbon concentration, expressed in ppm, found in the benthic sediments. (After Marchand *et al.*, 1979)

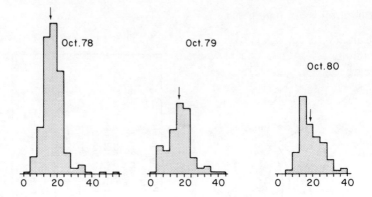

Fig. 3.52 Histogram of the frequency (as an absolute value from the sample taken from the total population) of the size of the *Patella* expressed in millimetres in a zone polluted by the oil-slick from the *Amoco Cadiz*. (After Raffin *et al.*, 1981)

One indirect effect which was unnoticed until recently but is now of grave concern, is the fact that during the incubation period of their eggs, seabirds contaminated by petroleum, however mildly, can pass on the pollution to their young. The result is a considerable embryo mortality rate and malformations in young birds hatched from eggs that have been subjected to external contamination. Albers (1977, 1978) provided experimental evidence of these effects which can occur with concentrations as weak as 5 µl/egg in mallard ducks and 20 µl/egg in eider ducks (*Somateria mollissima*). The earliest embryonic stages are the most sensitive.

Other research has shown that if mallard ducks (*Anas platyrhynchos*) are fed with crayfish contaminated with naphthalene this aromatic hydrocarbon is accumulated in the bird's organism (*in* Stickel *et al.*, 1979).

Some oil-tanker disasters have caused irreparable losses in colonies of marine birds. The accident of the *Gerd Maersk* in the Elba estuary in 1955 was responsible for the deaths of some 250 000 to 500 000 black scoter ducks (*Melanitta fusca*) alone. The case of the depopulation of the puffin colony (*Fratercula arctica*) in the Isles of Scilly, where the *Torrey Canyon* went aground, was particularly spectacular. The colony had more than 100 000 birds in 1907 but there were only about 100 left in 1967 after the tanker disaster. Another puffin colony, in the Sept-Îles, which had also been affected by the *Torrey Canyon* accident, was down to about 100 birds after the *Amoco Cadiz* oil-spill, as against 6000 birds counted in 1966.

More recently, a previously unknown cause of bird mortality linked to the petroleum industry has been discovered (Flickinger, 1981). This study showed that many birds die in the areas round oil-wells and petrochemical plants in Texas. The birds get stuck in the open waste-pipes of the discharge pits from disused oil-wells. It also happens in zones of industrial wasteland where petrochemical plants dump the waste from polystyrene manufacture and other petroleum by-products.

2. Pollution from detergents

There are several chemical groups of detergents. They can be divided more simply into anionic, non-ionic and cationic detergents according to the part of the molecule that has the surface-active properties.

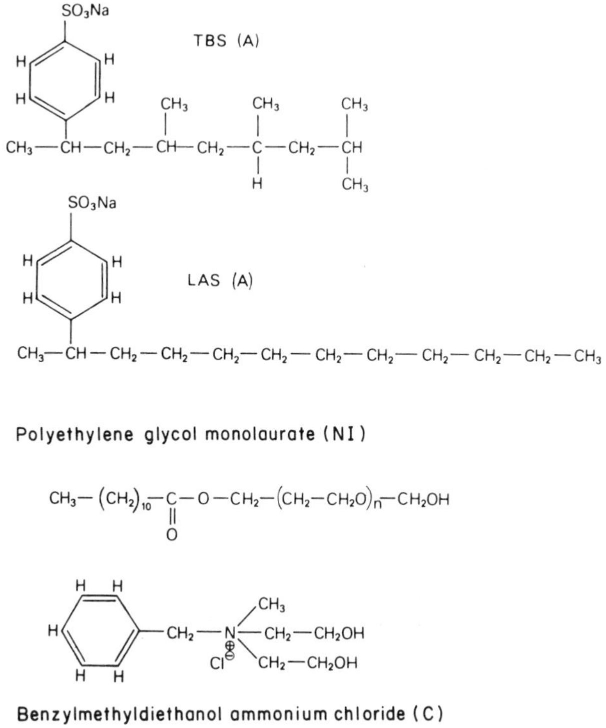

Fig. 3.53 Examples of the molecular structure of some detergents. A = anionic, NI = non-ionic, C = cationic

Anionic detergents

These are by far the most widely used, especially for domestic purposes, although certain cationic detergents are as important for industrial applications.

Among the anionic detergents there is a whole series of non-biodegradable compounds of which the best-known is tetrapropylene benzene sulphonate (TBS). In the early 70s a law was passed in France obliging manufacturers of commercial detergents to make their products a mix of at least 80% biodegradable detergents. These are generally anionics from the group of linear alkyl benzene sulphonates (LAS).

There are two principal causes of marine pollution by detergents: the discharge into the sea of urban and industrial waste contaminated with these substances, either directly from coastal towns and cities, or indirectly, carried by the rivers from inland towns. It is worth noting that no town or city on the French Mediterranean coast yet has a sewage treatment plant. A conglomeration the size of Marseille, with more than 1.5 million inhabitants, discharges its waste waters straight into the sea without any previous treatment and right in the middle of the famous Calanques coast.

Bellan and Peres (1970) found some 4 mg/l of detergents at the mouth of Marseille's main sewer and 0.1 mg/l about 10 km out to sea from the outlet. They also found 1 mg/l at the mouth of the Lez River in Languedoc which goes through Montpellier. An indication of the present detergent pollution levels was that nearly 1 mg/l of them were found in the water contained in benthic silts sampled from the underwater depths of the Golfe du Lion about 50 km out from the Rhône delta (Bellan, 1975).

The second cause of detergent pollution is their use to disperse or emulsify petroleum, particularly for dealing with the look (though not the effects) of oil slicks. The *Torrey Canyon* disaster, when more than 10 000 tonnes of detergents were sprayed into the sea near the British coast in a vain attempt to lessen the consequences of the hydrocarbon pollution, brought to light the ecotoxicological action of detergents in the marine environment.

Effects of anionic and non-ionic detergents—The *Torrey Canyon* disaster showed in fact that the detergents did even more damage than the petroleum to the invertebrate populations of the infralittoral and benthic zones. No single gastropod (patellae (limpets) Littorina (periwinkles)) and no lamelli branch from the intertidal zone survived. The crustaceans, the floating or burrowing Annelida and the echinoderms also perished in large numbers due to the detergents.

Bellan *et al.* (1972) made a detailed ecotoxicological study of the effects of some 50 different detergents on 9 species of Mediterranean invertebrates. The main results of their investigations are shown in Table 3.17. They show that the polychaetes are particularly sensitive to detergents, followed in order of

Table 3.17 The LC_{50} of anionic and non-ionic detergents for some marine invertebrates. (After Bellan *et al.*, 1972. *Reproduced by permission of Pergamon Journals Ltd.*)

Species	Detergents Anionic		Non-ionic	
	LC_{50} after 48 h	LC_{50} after 96 h	LC_{50} after 48 h	LC_{50} after 96 h
Polychaeta				
Capitella capitata	1 to 10 mg/l	1 to 5 mg/l	1 to 5 mg/l	0.1 to 2.5 mg/l
Scolelepis fuliginosa	10 to 25	0.1 to 25	0.5 to 5	0.1 to 2.5
Lamellibranchs				
Mytilus galloprovincialis				
from pure water	>800	5 to 25	1 to 25	0.5 to 5
from polluted water		5 to 25	1 to 25	0.5 to 5
Isopod crustaceans				
Sphaeroma serratum	>800	10 to 800	10 to 100	5 to 50

tolerance by mussels (*M. galloprovincialis*) and then isopod crustaceans (*Sphaeroma*). It also appears that non-ionic detergents are in general twice as toxic as the anionics. Also, their LC_{50} after 96 hours is significantly less than after 24 hours which proves their toxicity to be more persistent in water than with anionic detergents.

Cationic detergents

These are fortunately little-used as they are even more toxic than the other types to marine organisms. The LC_{50} after 48 hours for the isopod *Sphaeroma serratum*, which is fairly tolerant of detergents, is only just between 0.1 and 0.5 mg/l.

Long-term poisoning from even weak concentrations of detergents in water causes serious physiological disorders in invertebrates. Polychaete tubicolous worms cannot secrete their tube when concentrations reach between 0.5 and 100 ppm according to the detergent used. Bellan *et al.* (1971) found that just 0.01 ppm of the polyethylene glycol ester, a non-ionic detergent with very weak intrinsic toxicity, was sufficient to slow the growth and significantly reduce the fertility of the polychaete *Capitella capitata*.

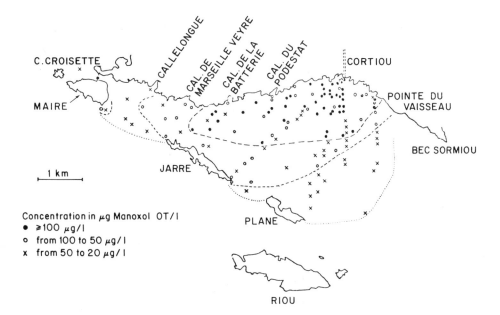

Fig. 3.54 Correlation between the amount of detergents in the sea and the scale of the regression of a population of *Cystoseira stricta* between Marseille and Cassis. The zone in which the population is changed has clearly been extended between 1965 and 1970. ––––– marks the polluted zone in 1965, ------- shows its extension in 1968, ... is the zone boundary in 1970, CAL = calanque. (After Arnoux and Bellan-Santini, 1972. *Reproduced by permission of the Centre d'Océanologie de Marseille*)

Detergents can also be associated with profound biocoenotic perturbations. Arnoux and Bellan-Santini (1972), for example, found a strong correlation between the anionic detergent concentration in the coastal waters of the Marseille region and the disappearance or significant regression of a population of a large freshwater phaeophyceaen, *Cystoseira stricta*, and the whole attendant zoocoenosis.

When detergents are used to clear crude oil from the beaches, it can have catastrophic consequences for the interstitial fauna. After the *Torrey Canyon* accident, the entire animal population of the polluted beaches was destroyed, with the exception of a few nematodes, some small oligochaetes and one type of spionid polychaete, all particularly tolerant of the detergents.

It does seem that the particular physical structure of the sandy substrates reduces the effects of the petroleum dispersants by absorbing them. However, concentrations of 100 ppm of detergent, as generally used to break down oil slicks, are capable of killing the entire meiofauna after only one week.

Bleakley and Boaden (1974) showed that the vitality and survival of some representatives of the shoreline fauna, the archiannelid *Protodriloides symbioticus* and some Harpacticoida, were greatly affected by 15 days of contact with concentrations equal to or above 10 ppm. Also, one year after certain experimental areas of beach were treated with 1 litre of detergents per square

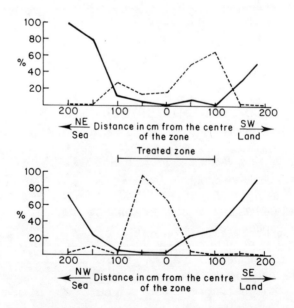

Fig. 3.55 The effects of a detergent used to disperse petroleum on the interstitial fauna of a beach on the Irish Sea. The percentage of survival (continuous line) and the mortality rate (dotted line) were measured 28 days after the start of contamination. (After Bleakley and Boaden, 1974. *Reproduced by permission of l'Institut Océanographique de France*)

metre, virtually the whole meiofauna was still missing. This slow reconstitution results from the detergents' persistence in the sand (cf. Fig. 3.55).

3. Pollution from solid wastes

The oceans are the place where all the waste products of our technological civilization either accumulate or are dumped. The most diverse objects, especially those almost indestructible ones made of plastic, are thrown into the sea and settle on the sea-bed near to the coastline. Some of them float, getting carried by the ocean currents to the most remote pelagic regions even in the Antarctic ocean (Gregory et al., 1985) which they then contaminate. This has been proved by samples gathered with neuston nets during various oceanographic surveys.

Many researchers have pointed out the virtual disappearance of the Mediterranean coastal sea-bed fauna and flora, one of the world's most polluted shorelines. Pollution by solid wastes from industry can be an important local problem. Not long ago, public opinion was roused over the dumping of yellow sludges in the Baie de Seine and red sludges in the Mediterranean. These sludges consist of powdery waste materials generally mixed with fresh water. Just off the French coast, near Cassis, red alkaline sludges, originating from an alumina production plant, are routinely dumped in an underwater canyon situated out to sea at a depth of 350 m. The sludges have been dumped at a rate of 85 m^3/hour since the spring of 1967. A regular study of the dumping and its consequences has been made and two zones identified in the form of dejection cones, stretched out lengthways. The first, an axial zone, is absolutely, azoïc and corresponds to the canyon floor, progressively being filled with the sludges, in which are found the shells of the species living in the bathyal silts. The other zone, outside the first, and called the deposit zone, is the subject of slower sedimentation. The zoocoenosis in this zone has been changed. It contains a certain number of the characteristic species from the bathyal silts that have survived the pollution (*Abra longicallus, Golfingia minuta* for example), species living in the sludges (*Lumbriconereis fragilis, Nucula sulcata*) and various ubiquitous species. The limivores seem to ingest without any harm the sediments mixed with red sludges and some polychaetes use them to build their tubes (*in* Peres and Bellan, 1972).

The red acid sludges discharged into the Tyrrhenian Sea off Livorno by the Montedison company seem to be far more poisonous. They are the residues from the manufacture of titanium dioxide and contain a mixture of sulphuric acid and ferrous sulphate with significant amounts of chromium, vanadium and other residual toxic metals. This dumping takes place in the middle of an 'upwelling' zone on the continental shelf between Cape Corsica and the Italian coast. It comes out on the surface and into a zone of shallow water no more than 180 m deep. All the planktonic life in the surface waters has been destroyed. Viale (1974) linked this pollution by red sludges with the numerous beachings of whales and other cetaceans in this part of the Mediterranean, most of whom had cutaneous damage. In a later study, Viale (1976) presented direct proof of the effects of this pollution on the cetaceans. An abnormal amount of titanium was found in the skin of the beached animals. The whole practice of this dumping threatens marine food chains involving species of economic value with scandalous contamination due to the considerable quantities of heavy metals and other exotic toxic elements contained in the red sludges.

II — Pollution of Limnic Ecosystems

1. Waste from the mining industries

Its Nature

Mineral- and coal-mines and various other ore-extractive industries discharge into the surface waters a certain amount of solid waste and the drainage water from their galleries. The result is contamination of the rivers and lakes receiving the effluents in the form of acids and various mineral salts. The latter come mainly from the oxidation of pyrites by physico-chemical factors or by bacteria.

Apart from the sulphuric acid formed, the waste waters contain various sulphates, especially of iron, aluminium, magnesium and calcium resulting from the reaction of the sulphuric acid (H_2SO_4) with the metals contained in the waste minerals. The waters also carry ferric oxide in suspension (ochre) and weak concentrations of copper, nickel, zinc, etc. — all highly toxic elements.

The dumping of waste mining products into continental waters is considerable. The largest American iron-ore mine was discharging some 220 tonnes of acid and sulphate waste per day into Lake Superior in 1974. In France, the potash mines of Alsace pour out annually into the Rhine 5 400 000 tonnes of sodium chloride in the form of concentrated brines. The accompanying pollution of the ground water (400 000 tonnes of salt per year) is affecting Colmar's water supplies. The town's drinking-water already contains 180 to 200 mg/l whereas the minimum that can be detected, from an organoleptic point of view, is 250 mg/l. Elsewhere, this pollution has caused the decay of more than a thousand hectares of forests on the plain of Alsace while the excessive salinity of the water prevents any new industrial development in the region.

Its Consequences

The contamination of limnic ecosystems by waste products from pyrite mines causes numerous problems. The ochre formed has a considerable spreading capacity, even in a relatively low concentration of 1 g/litre. The result is aesthetically disruptive as the ochre accumulates on the river beds and banks, colours them and clouds the water with its characteristic tint.

Generally associated with this type of effluent is acidification of the water, reducing its potential use. Even without the acidification, sulphate contamination of water can mean the water is no longer fit for numerous industrial purposes nor for consumption, particularly due to excessive magnesium content. The presence of iron, aluminium and magnesium in mine waste has damaging effects on all aquatic biocoenoses. The consequences of acidification for phyto- and zooplankton were discussed in the section dealing with sulphur dioxide pollution.

Added to this acidification is the toxic influence of the cations mentioned earlier. A concentration of iron as weak as 3 mg/l is enough to prevent maturation of the gonads and slow down the growth of *Gammarus minor* (*in* Glover, 1975). Aluminium in solution or in the form of neoformed suspension is particularly toxic: a concentration of 1.5 mg/l is fatal to the common trout (*Salmo fario*). The different cations can also greatly reduce the growth rate of freshwater Teleostei, especially the Salmonidae.

2. Pollution from lead shot

The hunting and shooting of waterfowl can contaminate the marshes and other humid zones with a significant amount of shot made of alloys of lead and antimony, both of which are highly toxic elements. On just one day of the French open season if all the 2.4 million French hunters (an excessive number considering the sustainable yield of the game population) shoot just one cartridge loaded with 32 g of lead, this amounts to a total of 76.8 tonnes of lead discharged into the environment. The shot, due to its almost total insolubility, will accumulate at the bottom of streams and rivers, etc., the habitats of freshwater vertebrates and invertebrates. It also constitutes a mortal danger to many water birds, particularly the Anseriformes — ducks, swans and barnacle geese, etc. — which like all birds, swallow pebbles and various hard objects that accumulate in their gizzard to aid the crushing of their food.

The birds that ingest this lead shot will rapidly suffer from lead-poisoning. A study of 35 500 Anatidae caught and X-rayed at the biological station of the Tour du Valat (Camargue) showed that 40% of the mallard ducks and 5% of the common teals (*Anas crecca*) were suffering more or less seriously from lead-poisoning (Hovette, 1972).

There is a high mortality rate among the contaminated bird populations. Davant *et al.* (1975) found swans that had died from lead-poisoning in the Teich bird reserve (Gironde) and who only had a few pieces of lead shot in their gizzard. With some dabbling ducks from the Fuligulinae group it only takes two or three pieces of lead shot to kill them.

Similar effects have been observed among the bird population of North America. In the Rice Lake (Illinois) reserve alone, Anderson (1975) noted the deaths of 1500 migrating Anatidae in the spring of 1972. Out of 96 *Aythya affinis* analysed, 75% had at least one piece of shot in their gizzard and 36% had more than 10 pieces. These dead birds had more than 46 ppm of lead in their liver.

Investigations concerning a related species (*Aythya valisineria*) (canvasback ducks) in Wisconsin and on the north-east coast of the United States showed that lead concentrations to the order of 0.2 ppm in the blood of these dabbling ducks were sufficient to inhibit the δ-aminolevulinic-acid deshydratase (ALAD) to 75%. 400 of the birds were watched for 3 years in the wild by a programme of capture and recapture (Dieter, 1979). The ducks which showed a reduction in ALAD activity contained in their serum average lead levels that were double those measured in birds with normal enzyme activity. Dieter suggests, therefore, using this enzymatic test to evaluate the degree of lead-poisoning caused by lead shot in wildfowl populations.

3. Pollution by detergents

Synthetic detergents are complex mixtures containing, as well as the active matter with its surface-active properties (cf. earlier section), a number of polyphosphates and several other ingredients: perborates, persulphates, 'bleaching' agents, perfumes, etc.

Causes of pollution

Not only do detergents pollute continental waters with their surface-active agents, but they also introduce a significant amount of phosphates into the waters. This unnatural increase in phosphorated compounds, though they are not toxic, can cause serious ecological damage to lakes, called dystrophication, which cannot be described in full here.

Most detergents in current use belong either to the anionic or non-ionic groups. Cationic detergents are prohibited from domestic use but have numerous industrial applications. The fact that most European countries have imposed a legal obligation on detergents to be biodegradable since the end of the 1960s has led to a reduction in the detergent pollution of rivers and streams. At the present time, surface waters generally contain less than 0.1 mg/l of detergents although some rivers may have concentrations of over 1 mg/l.

Their Effects

Detergents exercise various damaging effects in the limnic environment on the micro-organisms, plants and all the zoocoenoses. Their presence slows down the reoxygenation of the water, which is of particular concern as the high organic pollution levels in most of the rivers in industrialized countries are already depleting the amount of dissolved oxygen available.

The toxicity for living beings of some detergents, or their products after biodegradation, is not insignificant either. Many detergents are bacteriostatic and hinder the action of auto-purifying bacteria in water both in limnic ecosystems and in bacterial beds in water treatment plants. Freshwater animals can be poisoned by the levels of detergent concentration already found in some continental waters. For various limnic invertebrates, especially cladoceran crustaceans, copepods, gammarids and pulmonate gastropods, the LC_{50} after 96 hours is less than 1 ppm with certain surface-active substances (particularly cationic detergents).

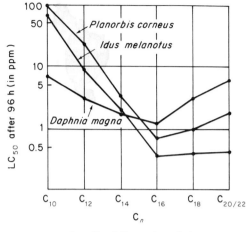

Fig. 3.56 The influence of the length of the carbon chain on the toxicity of cationic detergents (trimethylammonium chlorides in C_n) for various aquatic animals. (After Knauf, 1973. *Reproduced by permission of Carl Hanser Verlag*)

It has also been shown that the length of the aliphatic chain influences the toxicity of the detergent molecule. Knauf (1973) studied various alkyl sulphates and cationic detergent derivatives of trimethylammonium chloride. For these compounds, the LC_{50} after 96 hours is maximal for a C_{10} carbon chain and minimal (= maximum toxicity) for a C_{16} carbon chain.

As a general rule, the toxicity of detergents for freshwater fauna increases with the temperature, thereby combining its effects with the reduced levels of dissolved oxygen.

4. Pollution of continental waters by various organic substances

The modern chemical industry also discharges into continental waters a large variety of organic compounds. Among them are the phenols (petrochemical residues) and cyanides (metallurgy) whose incredible toxicity for all limnic fauna raises serious ecotoxicological problems.

There is not room in the present study to discuss the effects of these substances in detail. However, it is worth noting that some organic substances that could be generally regarded as inoffensive can also have disastrous ecotoxicological consequences. One example is that of the silicone oils (organosilexanes) which until recently were not recognized as an important potential menace to continental waters (Van der Post 1979). They are not in fact biodegradable in the secondary and tertiary treatment processes of polluted urban or industrial effluents. They codistil with water vapour and have no phase separation; they cannot be retained by the active carbons. Also, they are not oxidized by the bichromate treatment. Even more worrying is that trimethylsilanate inhibits the growth of the algae *Scenedesmus* sp. in concentrations of less than 2.5 ppm in water, thus affecting the functioning of lagoon and of activated sludge wastewater treatment plants.

CHAPTER FOUR

Nuclear Pollution

A — Concepts of radiobiology 205
  I — The ecological importance
    of different radioisotopes 206
    1. Principal types of
       radionuclides 206
    2. Ways of
       contamination 206
    3. Radiobiological units 208
  II — Biological effects of
    ionizing radiation 209
    1. The radiosensitivity
       of living beings to
       lethal doses 210
    2. Radiosensitivity to
       sublethal doses 211
B — Ecological consequences of
  radioactive fallout 214
    1. Contamination of
       the atmosphere 215
    2. Fallout 216
    3. Contamination of
       terrestrial ecosystems 216
    4. Contamination of
       the oceans 219
C — Radioecological consequences
  of the development of the
  nuclear industry 220
  I — Revision: the fuel cycle 220
  II — Pollution from nuclear
    power plants 221
    1. Causes of radioactive
       pollution 221
    2. The nature and
       quantity of waste
       products 222
  III — Pollution from spent
    nuclear fuel reprocessing
    plants 224
    1. Nature and
       importance 224
    2. The problem of
       radioactive waste 225
  IV — Contamination of
    trophic webs by the
    radionuclides discharged
    into the environment by the
    nuclear industry 226
    1. Contamination of
       limnic and marine
       ecosystems 227
    2. Contamination of
       agroecosystems 231
    3. The problem of
       external irradiation 233

The nuclear industry is today undergoing an extraordinary expansion. In the past no other new technology has grown so quickly if one considers the scale of this development. Huge markets are opening all over the world in a bid to cater to the energy 'needs' or rather energy greed of the industrialized nations.

The creeping petrol crisis and foreseeable shortage of fossil hydrocarbons in the not-too-distant future have greatly accelerated the nuclear power programmes in western countries. Although several other alternative energy sources could have been used to satisfy past demands, Europe has for the last 40 years or more based its energy supplies on that rare and badly-distributed resource—petroleum. Even the United States will have to import large quantities of crude oil from the Middle East by 1990 if its energy needs grow in the same way as they did up to 1973.

Predictions made in 1970 show that by the year 2015, installed nuclear capacity could have increased from 25 GW to 10 000 GW world-wide. In France alone, installed nuclear capacity could have increased from 3.5 GW to 170 GW electricity by the year 2000 (according to a government programme launched at the end of 1973 and now under construction), in other words it will go up 50 times in the space of 25 years, or double every 4 years. According to official plans, nuclear energy will account for one-half of all energy supplies estimated at 650 megatec/year (1 megatec = 1 million tonnes of coal equivalent) by the year 2000.

There are many causes for concern in these predictions. Firstly, they raise the question of the pollution potential of the nuclear industry given the levels of activity it will soon be reaching, along with the question of industrial use

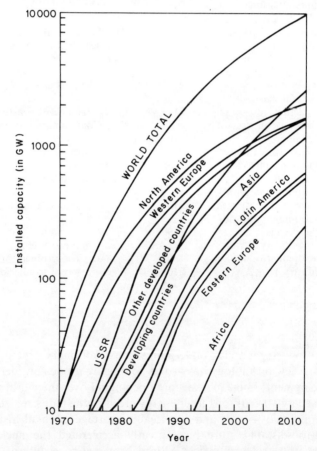

Fig. 4.1 Predictions for the growth of installed nuclear power capacity between 1970 and 2015 (according to the IAEA, 1971.) (*Reproduced by permission of the International Atomic Energy Agency, Vienna*)

of zones previously given over to agriculture or fishing. Among the numerous ecotoxicological problems associated with nuclear energy are the following: Do we know with any certainty what the chronic or long-term biological effects of radiation may be? Have all the necessarily important criteria of environmental protection been taken into consideration at every stage of the nuclear fuel cycle? Are the ecological risks associated with the problem of waste being underestimated, particularly those concerned with discharge into continental or marine waters?

It is true that the problem of nuclear pollution is not new and that consequently the peaceful applications of nuclear energy have not developed without the consideration of radioecological factors. There has been a considerable accumulation of knowledge on the subject over the last 40 years or more since the American decision in 1941 to develop nuclear weapons. The Hiroshima and Nagasaki bombs and frequent atmospheric tests on so-called deterrent weapons have all been the source of significant radioactive pollution giving rise to numerous ecological studies.

It is necessary at this point to look again at the fundamental concepts regarding the biological effects of radiation.

A—CONCEPTS OF RADIOBIOLOGY

Radioactivity involves several types of rays: α rays, ionized helium nuclei with very weak penetrating power which are stopped by the outer layers of the skin; β rays which can go through a few cm thickness of tissues or even a few dm with the most penetrating of these rays; and lastly the γ rays, electromagnetic in nature and, therefore, comparable to X-rays, and capable on occasions of going through lead screens that are several metres thick.

All of these radiations are called ionizing because they are able to strip the electrons from the outer layers of atoms, and therefore to ionize them. The ions produced in this way are very reactive chemically and can modify various cellular constituents, causing, for example, peroxides and other cytotoxic compounds to form. A stronger dose of radiation, producing numerous ions will sooner or later bring about the death of the exposed cells. Even lower doses, with no apparent damage resulting, can still induce irreversible structural modifications to the DNA (mutations).

Since its earliest beginnings, life on Earth has had to adapt to natural radiation in the environment, especially when moving from the ocean to the land which does not offer the same protection against external radiation as a screen of water. A certain amount of radiation does in fact exist in natural biotopes. It is either of a cosmic nature or emitted by natural radioactive compounds contained in the Earth's crust: uranium, radium, thorium, as well as various radioisotopes from biogenic elements (^{40}K or ^{14}C, for example). A total of about 50 natural radioisotopes have been counted in the biosphere. Like all radioactive elements, they disintegrate spontaneously: their mass decreases in exponential manner as a function of time. The time needed for the mass of a radioelement to decay by half is called the **half-life**. This can be very variable, ranging from a fraction of a second for the most unstable radioelements to several *billion* years (4.5 x 10^9 years for ^{238}U).

The ecological consequences of this property is that the only way to make radioactivity disappear is to give the radioisotope time to disintegrate spontaneously. In practice, fighting nuclear pollution can never be other than preventive as there is no possibility of biodegradation or any other mechanism that could eliminate this type of contamination from the natural environment.

I — The Ecological Importance of Different Radioisotopes

1. Principal types of radionuclides

They are not all of the same importance. Radionuclides with a short half-life, of 2 days for example, will not be the most dangerous as they disappear quickly from the environment after contamination occurs. At the opposite end of the scale, substances with a very long half-life will also be virtually harmless because they emit a very low number of rays per unit of time.

The most dangerous radioelements will, therefore, be those with an intermediate half-life, to the order of a week, a month or a year as they will have the time to accumulate in different organisms and in the food webs (cf. Chapter 2, page 84). In addition, the radioisotopes of fundamental constituents of living matter (^{14}C, ^{32}P, ^{45}Ca, for instance), given the same level of contamination, will be far more dangerous for living beings than other non-biogenic elements which are rarely or never absorbed by organisms.

Another category of radioelements that presents formidable ecotoxicological properties is the one that contains substances with a chemical character similar to that of the fundamental constituents of living matter. The similarity between the chemical properties of strontium and calcium or between caesium and potassium mean that both radiostrontium and radiocaesium are particularly dangerous. ^{90}Sr is easily incorporated into the skeleton of vertebrates due to its closeness to calcium, and ^{137}Cs can accumulate in the muscles just like potassium. As the half-life of these two radioelements is about 30 years, they are able to exercise their damaging effects over the whole lifespan of a contaminated human being.

2. Ways of contamination

Another basic difference exists in radiobiology between the ways of radiation, linked in part to the type of radioelement concerned. There is **external radiation**, caused by rays present in the environment, and **internal radiation** following the inhalation or ingestion of radionuclides.

The absorption of biogenic radioelements or radionuclides with similar chemical properties is by far the most dangerous. With all due allowance being made, inhalation of ^{85}Kr, a rare chemically inert radioactive gas, or ingestion

Table 4.1 The principal radioelements of ecological importance.
Group A: Radioisotopes of the fundamental constituents of living matter.
Group B: Fission and neutron-capture products formed during nuclear explosions or inside nuclear reactors.
Group C: Rare radioactive gases formed under the same conditions as in B

Group	Radioelements	Half-life	α	β	γ
A	Carbon (^{14}C)	5 568 years		+	
	Tritium (^{3}H)	12.4 years		+	
	Phosphorus (^{32}P)	14.5 days		+++	
	Sulphur(^{35}S)	87.1 days		+	
	Calcium (^{45}Ca)	100 days		++	
	Sodium (^{24}Na)	15 hours		+++	+++
	Potassium (^{42}K)	12.4 hours		+++	++
	Potassium (^{40}K)	1.3×10^9 years		++	+
	Iron (^{59}Fe)	45 days		++	+++
	Manganese (^{54}Mn)	300 days		++	++
	odine (^{131}I)	8 days		++	++
	Cobalt (^{60}Co)			+++	+++
B	Strontium (^{90}Sr)	27.7 years		++	
	Caesium (^{137}Cs)	32 years		++	+
	Cerium (^{144}Ce)	285 days		++	+
	Ruthenium (^{106}Ru)	1 year		+	
	Yttrium (^{91}Yt)	61 days		+++	++
	Plutonium (^{139}Pu)	24 000 years	++++		++
C	Argon (^{41}Ar)	2 hours		++	
	Krypton (^{85}Kr)	10 years		+	
	Xenon (^{133}Xe)	5 days		+++	

of thorium, an actinide of low atomic weight rarely penetrating the intestinal barrier, is less dangerous than inhaling ^{14}Co$_2$ or ingesting radioiodine which fixes itself in the thyroid gland after just a few minutes.

However, in the case of sudden or accidental ingestion of a radionuclide it is not true that the only possibility for decontamination of the organism is to wait for the radioactive disintegration of the element. Every simple element has in fact a biological half-life (T_b) of its own, linked to its biological turnover: the intake of carbon, nitrogen, phosphorus, etc., compensates precisely in an adult for the losses caused by respiration and excretion. If T_p is the physical half-life for a radionuclide, the effective half-life, T_e, of the element in the given organism will be given by the equation:

$$\frac{1}{T_e} = \frac{1}{T_p} + \frac{1}{T_b} \qquad (1)$$

or

$$T_e = \frac{T_b \cdot T_p}{T_b + T_p} \qquad (2)$$

The total radioactive activity of a contaminated organism for a time t, A_t, is given by the expression:

$$A_t = ae^{-kt} \qquad (3)$$

where a is the activity at zero time, which depends on the initial quantity ingested of the radioelement and its half-life T_p, and k represents a constant of elimination particular to the organism being studied and linked to the biological half-life of the element by the equation:

$$k = \frac{\log 2}{T_b} \qquad (4)$$

However, even an element with no biogenic properties such as ^{85}Kr or plutonium can cause radioecological problems if released in significant quantity into the environment, as is the case with the present high growth-rate in the nuclear industry. Although krypton-85 is in fact a rare gas with, therefore, no biological activity, its very chemical inertia enables it to accumulate in the atmosphere where it could cause serious climatic disturbances (see page 234).

3. Radiobiological units

In order to be able to compare the effects of ionizing radiation on individuals, species and biocoenoses, it is necessary to have units with which the degree of contamination of different communities and the quantity of radiation received by each organism can be measured.

The oldest unit of radiation is the **curie** (Ci). This corresponds to the disintegration of 3.7×10^{10} atoms per second, that is with fairly good approximation the quantity of radiation emitted by 1 gram of radium per second. This is a very considerable quantity. Radioecologists often have to use, therefore, the microcurie (μCi) which is 10^{-6} Ci, the nanocurie (nCi) (or millimicrocurie) which is 10^{-9} Ci and the picocurie (pCi) (formerly called the micromicrocurie) which is 10^{-12} Ci. A new unit, the Becquerel is currently in use for measuring environmental radioactivity (1 Bq = 27 pCi).

The **rad** is the basic unit of radiobiology. It is defined as the radiation dose corresponding to the energy absorption of 100 ergs per gram of tissue of the affected organism.

The **rem**, a unit derived from the rad, takes into account a correction factor called the relative biological effectiveness (RBE). This allows for the fact that absorption of ionizing radiation by living matter depends on the nature of the radiation. α, β, γ or X-rays and neutrons do not have the same effects from an equal amount of incident energy. The RBE of a thermal neutron will be 10 times higher than that of a γ ray or of an X-ray of 320 keV, with all else being equal. In cases where radiation is limited to an electomagnetic γ ray or to an X-ray of 320 keV, it can be said that 1 rad = 1 rem.

A new unit has recently been introduced into radiobiology, destined to replace the rem and used to measure absorbed radiation doses taking account of the biological effectiveness. It is called sievert (Sv). 1 sievert = 100 rems.

II — Biological Effects of Ionizing Radiation

Living cells are not all equally sensitive to radiation. As a general rule, prokaryotes are much more resistant than eukaryotes. If all else is equal, it is the cells with a high mitotic index that show the greatest sensitivity to radiation.

The basic biochemical mechanisms governing the action of ionizing radiation on the cellular level relate to their effects on the macromolecules of nucleic acids. As they induce chemical alterations to the DNA, it explains how radiation

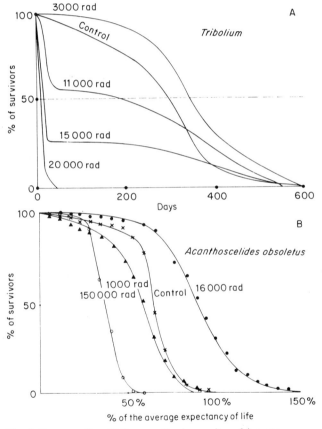

Fig. 4.2 The influence of radiation on the longevity of insects.
A: The survival graphs for a coleopteran of the genus *Tribolium* exposed to various doses.
B: The graphs for the bruchid *Acanthoscelides obsoletus*. In both cases it can be seen that the low doses signficantly increase the lifespan, which is the opposite to what is observed with warm-blooded animals. This phenomenon is the result of a slowing-down of the metabolism of the irradiated insects. (A after Cork, 1957. B after Echaubard *in* Ramade, 1982.)

can cause mutagenesis and carcinogenesis. Any modification to the *structure* of the bases leads to a mutation of the affected codon. Similarly, the pure and simple breaking of the polydesoxyribose phosphate skeleton causes chromosomal mutations to occur if the two chromatids are affected simultaneously.

1. The radiosensitivity of living beings to lethal doses

With strong doses of radiation causing quite rapidly the death of a number of the exposed organisms the dose–response curve is of the sigmoïd type with a threshold. There are great variations in the radiosensitivity of living beings according to their taxonomic position.

The LD_{50} following one dose of radiation is to the order of 1 million rads for bacteria, a few hundred thousand rads for green plants, about ten thousand rads for arthropods and only a few hundred rads for warm-blooded vertebrates. It appears, therefore, that the radiosensitivity of living species is greater the higher their degree of evolution and complexity of their organism.

However, there can be considerable variations between organisms within the same systematic group. The LD_{50} for the adult *Drosophila* (fruit fly) is 85 000 rads, while for the house-fly it is only 10 000 rads. Among the arthropods, scorpions and certain bruchid Coleoptera show exceptional resistance to radiation, with an LD_{50} to the order of 150 000 rads.

At the other end of the scale, birds and mammals are extremely sensitive to radiation. The LD_{50} for man is about 450 rads if the mortality is measured in the month following irradiation. Exposure of humans to 100 rads does not cause immediate death but a significant increase in the number of cases of cancer. This level of exposure also causes permanent sterility in women and 2 to 3 years of sterility in men.

The most sensitive tissues to radiation are those with the highest mitotic activity, so there is maximum radiosensitivity in the gonads, in embryonic cells and in the haematopoietic bone marrow (which explains the death of a high proportion of the Hiroshima and Nagasaki victims from leucopoenia).

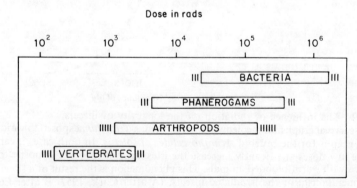

Fig. 4.3 The scale of radiosensitivity for the principal groups of living organisms

Table 4.2 The effects of ionizing radiation on different tissues of the rainbow trout *Salmo gairdneri*. (Rice and Wolfe *in* Hood et al., 1971, *Impingement of Man on the Ocean* Wiley Interscience)

Stage of life-cycle	LD_{50} (in reds)
Gametes	50 to 100
Egg — first segmentation division	58
Segmentation stage 32	313
Germinal disc	460
Optical vesicle (according to stage)	410 to 900
Adult	1500

2. Radiosensitivity to sublethal doses

Exposure to doses of radiation that do not cause immediate death can still trigger off a series of noxious biological effects:
— Irradiation reduces the physiological strength of exposed subjects. It decreases biotic potential, growth, resistance to toxicants and the immune defence system of affected organisms.
— Irradiation alters the genome in a delayed action. It causes damaging, sublethal mutations which come out in the second or third generations.
— Irradiation is cumulative, a build-up of irreversible effects. However, this accumulation is not absolute, due to repair mechanisms existing for the nucleic acids.

Effects of Radiation on the Longevity of Organisms

If one studies the effects of sublethal doses of radiation it becomes apparent that the life-span of individuals is affected even if there is no immediate lethal response to the radiation. In general, the longevity is reduced, although for certain insects weak doses of radiation can increase their average life-expectancy.

Epidemiological studies of human beings have shown that sublethal doses of radiation cause a highly significant increase in the incidence of cancers. The report of the American National Academy of Sciences (1972) fixed at 1% the rate of increase in cancers caused by irradiation of the whole American population by a dose of 0.1 rem/year (in addition to natural radiation). However, the last report of the same BEIR Committee of the American National Academy of Sciences (1980) is far less categorical in its conclusions than the earlier report. It stresses the methodological gaps in the field of evaluating the effects of doses of about 100 mrem/year. By adopting various hypothetical dose–response curves for low-level radiation, the later BEIR report fixes at 7% (in the worst possible case) the rate of increase in the incidence of cancers for a population exposed to a dose of 1 rem/year throughout their life. But this increase is incalculable because it is too near 0% in the dose–response curve model considered by the experts to be the most probable.

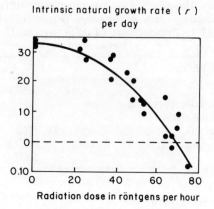

Fig. 4.4 Demoecological effects of continual exposure of a population of *Daphnia pulex* to low doses of X-rays. There is a negative correlation between the exposure level and the biotic potential. (After Marshall, 1962)

Effects on Biotic Potential

Sublethal doses also affect the biotic potential of irradiated organisms. Repeated or continual exposure to ionizing radiation causes a progressive reduction in the intrinsic natural growth coefficient (r). Marshall (1962) studied this phenomenon in detail in a population of *Daphnia pulex* exposed permanently to weak doses of X-rays. The resulting decline in biotic potential comes not only from the shortened breeding period due to the reduced lifespan of the females, but above all from the sterilizing action of the radiation on the gonads of both sexes.

There are also great variations in the radiosensitivity of germinal cells according to the species considered. For arthropods, radiosterilizing doses are between 1000 and 80 000 rads. They go up to 96 000 rads for the nematode *Ditylenchus dipsaci*, which is particularly resistant to radiation.

Mutagenic Effects

One of the most worrying consequences of exposure to low radiation levels is

Table 4.3 Doses causing 100% radiosterilization in various species of arthropods

Species	Dose sterilizing 100% of the specimens (in rads)
Icerya purchasi (Homoptera)	1 500
Musca domestica (Diptera)	4 000
Drosophila melanogaster (Diptera)	16 000
Habrobracon sp. (Hymenoptera)	7 500
Lyctus sp. (Coleoptera)	32 000
Tyroglyphus farinae (Acarina)	50 000

Table 4.4 Estimation of the mutation rate in the human species caused by various sources of radiation

Cause of mutation	Annual dose of radiation over 30 years (in rads)	Mutation rate caused out of 100 births
Natural radiation	3	2 %
Fallout from nuclear experiments	0.3	0.02 %
Medical radiography	3	0.2 %
Television	0.3	0.02 %
Maximum 'permissible' dose of radiation	5	0.33 %

their mutagenic power. It must also be remembered that these effects do not occur only in the exposed individual but also in his or her descendants. The mutation rate in the offspring of irradiated parents will increase proportionally to the dose they receive during their life. Under these conditions, even a minimal increase in the mutation rate, which would pass unnoticed in the first generation, could, if maintained permanently, lead to a genetic catastrophe in the unforeseeable future.

Various experts have calculated the increase in the natural mutation rate in the human species, resulting from the different technological sources of radiation (cf Table 4.4).

Over the last few years another course of investigation has been folowed, designed to evaluate the eventual mutagenic effects on humans resulting from low doses of radiation, by studying populations that live in regions with a very high level of natural radiation.

In several countries there are in fact high-grade uranium- and thorium-bearing ore deposits, sources of abnormally high background radiation. This is the case with Kerala in India, the province of Espirito santo in Brazil, and the region of Yangjang in the province of Guangdong, China. The absolute record is held by the uranium deposit at Lodève in the south of France where in a vineyard covering almost a hectare a radiation level of nearly 100 rem/year was found (Leonard et al., 1979).

Experiments carried out *in situ* at Lodève on rodents kept for several months in this highly-radioactive zone show chromosomal mutation, but epidemiological studies on autochthonous human groups in other regions are not conclusive. Kochupillai *et al.* (1976), studying a population in Kerala exposed to 3 rem/year, observed a significantly higher incidence of Down's syndrome compared with a control group living in a zone where the natural radiation is 100 mrem/year. However, the validity of the sample has since been questioned. Elsewhere, Lüxin and the Chinese study group into high levels of natural radiation (1980) have not managed to find evidence of an increase in the mutation rate among a population of 90 000 living in a region of the province of Guangdong where the upper layers of rock are rich in a thorium-bearing ore, monazite, and where

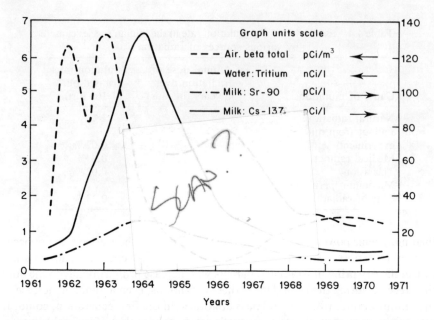

Fig. 4.5 The variation as a function of time of the intensity of radioactive fallout over the United States during the 1960s. After the treaty of 1962 prohibiting atom bomb tests, the result was a progressive reduction in fallout from 1965–66 onwards, proceeded by a decrease in the radioactivity of the air from 1963–64 onwards. (Source: US President's Council on Environmental Quality, Washington DC, 1971, the Mitre Corp)

the average level of irradiation of the population is three times higher than for the control population.

Consideration of the risks to human populations from radiation, particularly those of mutagenesis and carcinogenesis, have led the public health authorities to set standards for maximum 'tolerable' doses of radiation. These have been fixed at 5 rems/year for workers in the nuclear industry (called a high-risk group) and at 0.5 rems/year for 'isolated individuals' (that is, people living near nuclear installations). Finally, the maximum permissible dose for the rest of the human population is fixed at 0.17 rem/year. These doses are considered to be in addition to radiation from natural sources (cosmic rays, radioactive rocks, natural and artificial radioelements).

The results were published some years ago (Neel, 1981) of an epidemiological survey of the descendants of Hiroshima and Nagasaki victims over the 34 years following the irradiation of their parents. These results suggest that the dose which doubles the mutation rate in human beings would be 156 rems or four times higher than the dose established by experiments on rodents.

B—ECOLOGICAL CONSEQUENCES OF RADIOACTIVE FALLOUT

1. Contamination of the atmosphere

The problem of radioactive pollution was evident long before the peaceful development of nuclear power. By 1962, the year when the treaty was signed prohibiting atmospheric nuclear tests, the fission products from some 170 megatons had already contaminated the ecosphere, which is the equivalent of about 8500 Hiroshima-type bombs. To this must be added the rare radioactive gases and tritium emitted by nuclear fuel reprocessing plants. However, this does not take into account the fact that radioelements are liable to cause fallout. During the period of atomic bomb tests, 1954–62, 9 megacuries of ^{90}Sr and 14 megacuries of ^{137}Cs were released into the atmosphere. This would give 24 millicuries of ^{90}Sr and 39 millicuries of ^{137}Cs per square kilometre.

In reality, radioactive residues are deposited in a very irregular way over the surface of the continents and oceans. About 30% of the initial quantity produced falls near to the source of the explosion. The rest, which has gone into the troposphere or even the stratosphere, will only return very slowly to the Earth's surface. In Fig. 4.6 it can be seen that the fallout is maximal in the zone between the latitudes of 40 and 60 degrees North. In certain areas of the EEC in 1963, fallout reached some 2 curies per square kilometre, which was 50 times more than the amount calculated to fall if there had been an even distribution throughout the biosphere.

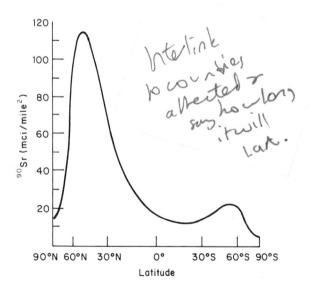

Fig. 4.6 Distribution by latitude of fallout of strontium-90 in 1964. Note the peak between 45 and 60 degrees North, but also the contamination of the southern hemisphere, where there were no atom bomb tests, as a result of the exchange of air masses between the two hemispheres. (After List et al., 1965)

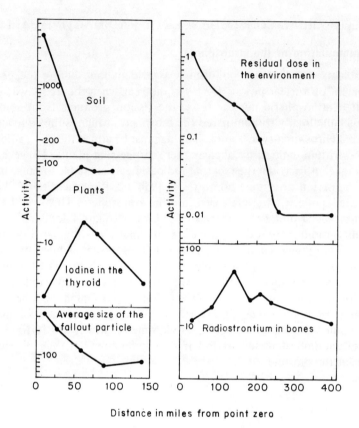

Fig. 4.7 The correlation between the distance from point zero (Nevada test site) and the strength of radioactive contamination of the biotopes and the biomass. (After Nishita and Larson *in* Odum, 1959)

2. Fallout

Precipitations bring down to Earth the radioactive waste matter that has been dispersed into the atmosphere. The fallout will then make its way into the soils and waters and eventually be incorporated into the biomass.

3. Contamination of terrestrial ecosystems

Incorporation into Trophic Chains

The retention of radioelements by soils makes the soils partially unfavourable to plants. However, a fraction of the substances will be absorbed by the plants through their roots. The plant biomass can also be contaminated through their leaves.

Moving up the trophic chains, the result will be a process of biological magnification of the radioactive pollution. Human food supplies will be contaminated to different degrees, sometimes seriously, as is the case with farm products through contamination by ^{90}Sr and ^{131}I of dairy products and ^{137}Cs in milk and meat. Although the major part of most radioisotopes is retained by primary producers and other lower levels in the food chains, and yet for all that significant amounts in absolute terms can reach the higher levels independently.

The following food chain has proved to be very vulnerable to pollution from radioactive fallout:

$$\text{soils} \rightarrow \text{pasture} \rightarrow \text{domestic animals} \begin{smallmatrix} \nearrow \text{milk} \searrow \\ \searrow \text{meat} \nearrow \end{smallmatrix} \text{man}$$

(a) Example of the Lapps—An earlier section described the contamination of the food of arctic populations. It is worth adding that lichens can absorb 95% of the amount of radioactive fallout that has contaminated them. As the reindeer (called caribou in Canada) feed almost entirely on lichens in winter, the trophic chain leading to the Lapps, Eskimos and various arctic populations in the north of the USSR can reach a high level of contamination, especially as these people live in latitudes where maximum fallout occurs. Given that the ration of reindeer meat can be up to 6 kg per week for an adult man, the bodily contamination level of these people can reach a concentration of ^{137}Cs or ^{90}Sr a hundred times higher than that of people living at lower latitudes.

(b) The case of radioiodine—There is a remarkable correlation between the level of radioactive pollution in the air, the date of atom bomb tests and the pollution levels of foodstuffs by radionuclides. The delay between the maximum level of air pollution and that of animal products (milk, for example) comes from the necessary time lapse before fallout reaches its maximum amount in the soil (cf. Fig. 4.5).

The absorption of radionuclides into food webs is very rapid. One of the most dangerous fission products, radioactive iodine, can make its way in just a few days into dairy products and from there into the thyroid gland of humans. Studies of the correlation between atmospheric nuclear bomb tests and contamination of the thyroid glands of cattle and humans by ^{131}I showed the great speed of this contamination. The thyroid follicles accumulate in their colloid most of the ^{131}I that reaches them in just a few hours, due to the highly selective permeability of the follicle cells and an extraordinarily efficient process of active transport.

The rapid absorption of radioiodine by the thyroid gland is a rapid and observable phenomenon that occurs even in areas situated far away from any nuclear arms test sites. For instance, there was a significant increase in the amount of iodine-131 in the milk and thyroid glands of cattle in France following a Chinese nuclear test during the period of 15 October to 15 December 1976 (Morre et al., 1977), whereas the previous levels had been virtually zero due to the Soviet–American agreement of 1963.

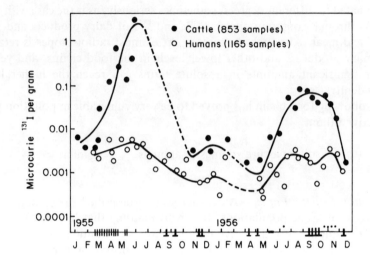

Fig. 4.8 Correlation between the dates of atmospheric nuclear bomb tests, and the levels of contamination by radioiodine of the thyroid glands of cattle and humans. Note the very slight difference between the peaks on these graphs for cattle and those for humans, showing the speed that ^{131}I enters the trophic chains. (After Comar et al., 1957, Science, **126**, 16–18. Copyright 1957 by the AAAS)

(c) The case of plutonium — Studies of contamination of the biosphere by plutonium show that both atmospheric and soil pollution are essentially the result of atmospheric atom bomb tests, and to a lesser extent from satellites containing this element falling back to Earth. After a rapid decline in the pollution level during the 1960s, due to the suspension of nuclear arms tests, there was a tendency towards stabilization during the 1970's, with the quantity of plutonium-239 and plutonium-240 in the stratosphere varying around an average of about 1500 curies (Harley, 1980).

Mathematical Models for Measuring Contamination

The movement of ^{90}Sr and ^{137}Cs has also been studied carefully in view of their high radiobiological toxicity. Robison and Wilson (1973) proposed the following formula for calculating the level of contamination of the human body by these dangerous radionuclides;

$$C_B = \frac{R(UAF)f_M I f_B}{\lambda_p m \lambda_B} \tag{5}$$

where

C_B = the concentration of the radionuclide in the reference organ of the human body (in µCi/g)
R = intensity of the fallout in µCi/m²/day
UAF = coefficient of the use of the area of pasture by the cow in m²/day
f_m = coefficient of transfer of the milk, or the amount of radionuclide ingested daily that is secreted in the milk (in day/litre)
I = average ingestion of dairy products in litres/day
f_B = the fraction of the quantity of radioisotope reaching the reference organ
λ_p = the effective elimination constant from the plant (per day)
m = mass of the organ in grams
λB = the effective elimination constant from the man (per day)

Knowing C_B, it is then possible to calculate the level of irradiation of the man expressed in rem/year (DR)

$$D_R = 1.85 \times 10^4 E\ C_B \tag{6}$$

where E = the effective disintegration energy (in MeV)

4. Contamination of the oceans

The world's oceans have also been greatly polluted by radioactive fallout. Evidence has been found of a significant increase in the concentration of ^{14}C in the first 300 metres below the surface. Concentrations of ^{137}Cs and ^{90}Sr are going up similarly.

It would seem that fallout affects mainly the upper layers of the oceans for the present. However, Broecker et al. (1966) found detectable concentrations of ^{90}Sr and ^{137}Cs in the ocean deeps of the Atlantic at a depth of 4330 m, proving that vertical circulation in the oceans is far more rapid than was previously thought. The greatest caution must be taken, therefore, with techniques for dumping nuclear waste in deep-water trenches.

The oceanic biomass is also contaminated by the fallout from nuclear tests. This pollution has spread several thousand kilometres away from the small islands where tests were held. The explosion of the first thermonuclear bomb in Eniwetok in 1953 led to pollution of the ichthyofauna throughout the extreme eastern part of the Pacific. Bonitos caught in Japanese waters were so radioactive that they were not fit for consumption. For several months all Japanese fisheries were checked with Geiger counters to ensure protection of the population from radiation, and fish with radiation levels of over 100 pulses per minute were immediately destroyed.

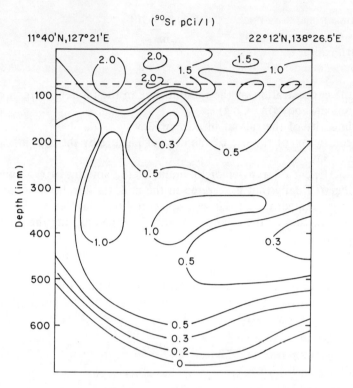

Fig. 4:9 The distribution of ^{90}Sr in the Pacific Ocean between 11 and 22 degrees North and 127 and 138 degrees East (transect passing through the Bikini atoll). The dotted line marks the position of the thermocline. These measurements, taken in March 1955 show a clear depth gradient with contamination detectable in the first 600 m depth. (From Miyake, 1971, in Hood *et al.*, *Impingement of Man on the Ocean*, Wiley Interscience)

C—RADIOECOLOGICAL CONSEQUENCES OF THE DEVELOPMENT OF THE NUCLEAR INDUSTRY

I—Revision: the fuel cycle

The nuclear power industry involves a succession of necessary phases making up the fuel cycle. It starts with extraction of the ore. The only natural fissile material, ^{235}U, is found in very small quantities (1 part per 140) in natural uranium whose main isotope is ^{238}U. The enrichment process, designed to raise the isotopic ratio of ^{235}U, is not completely necessary for reactors to function (cf. British and French natural uranium gas-cooled reactors or Canadian Candu reactors).

The fissile material is prepared in the form of 'cartridges' or 'rods' clad in alloys with suitable neutronic and thermal properties ('zircalloy', for example). All these 'fuel elements' form the 'core' of a nuclear reactor. There are usually about 40 000 of them in the case of a PWR reactor producing 920 MW of electric power.

A whole range of radioactive waste products is produced by a nuclear power-plant—rare gases are emitted into the atmosphere and diluted effluents from the purification of the cooling system are discharged into surface waters. However, the most polluting stage of the fuel cycle is not when it goes through the reactor, but later, from the moment when the spent fuel elements are removed from the reactor to be treated in specialized reprocessing plants. After remaining in the reactor for a certain time, and when at best 75% of the fissile material has disintegrated, the fuel elements have to be removed because they have been 'poisoned' by various fission products (neutron-capturing elements are formed which decrease the reactor's output, and the rate of non-fissile ^{236}U formation goes up, etc.), and also to extract from them the plutonium formed by capturing neutrons from ^{238}U. Plutonium, apart from its military uses, is an indispensable element for fast-breeder reactors.

The main pollution difficulties arise, therefore, in the spent fuel reprocessing plants (such as the La Hague type in France or Windscale in UK). Their function is to separate the residual fissile matter from the various radioactive 'wastes', which comprise a whole series of radionuclides:

— Fission products from the disintegration of the uranium atom, grouped round two peaks corresponding to the atomic mass around 90 (^{90}Sr, ^{85}Kr, for instance) and 130 (^{131}I, ^{133}Xe, ^{137}Cs, for example).

— Neutron capture products caused by irradiation from thermal neutrons of the different construction materials in the reactor (such as ^{54}Mn, ^{60}Co).

— Fissile materials mainly formed from unburned plutonium residues produced by neutron capture from ^{235}U and ^{238}U. In fact, as no recuperation technique is 100% efficient, the amount of plutonium remaining in treatment waste is estimated to be about 1% of the total. This waste also contains various transuranium elements (americium, curium, etc.), alpha emitters with long half-lives, like ^{239}Pu (half-life of 24 000 years).

II—Pollution From Nuclear Power Plants

1. Causes of radioactive pollution

A light-water reactor of the PWR type is an important source of both water- and air-pollution. Cladding cracks quite frequently, with the result that the water in the primary cooling circuit becomes charged with numerous fission products and activation products formed from irradiation of the reactor's metallic structures or from the impurities in the water. In particular, a significant quantity of tritium is formed by a ternary reaction from the boron contained in the primary circuit and also by the activation of the hydrogen and deuterium.

Contamination of the primary circuit in this type of reactor is aggravated by the fact that it is only recharged when the reactor is shut down, once a year. This means that when the cladding cracks there is intense pollution of the primary coolant as the faulty fuel elements cannot be removed from a reactor while it

Table 4.5 The principal radionuclides discharged into the environment by a light-water reactor of the PWR type of 1050 MW electricity. (After Rice and Wolfe *in* Hood, 1971)

	Liquid effluents				
Isotope	Half-life	Quantity in μc/year	Isotope	Half-life	Quantity in μc/year
^{3}H	12.26 years	4×10^9	^{133}I	21 hours	5.13×10^3
^{54}Mn	314 days	9.7×10^1	^{137}Cs	32 years	4.58×10^3
^{56}Mn	2.6 hours	2.64×10^1	^{140}Ba	12.8 days	2.28
^{60}Co	5.2 years	3.48	^{144}Ce	285 days	7.82
^{90}Sr	28 years	5.76			
^{91}Y	59 days	2.11×10^1			
^{99}Mo	60 hours	1.25×10^4			
^{131}I	8 days	6.61×10^5			

	Gaseous effluents	
Isotope	Half-life	Quantity in curies/year
^{85}Kr	10.4 years	5.62×10^3
^{133}Xe	5.27 days	1.58×10^3

is running. Consequently, in spite of the previous treatment of effluents from a primary circuit which is bound to retain most of its contaminating radionuclides, the waste water that is routinely dumped into the neighbouring river or sea is only partly purified.

2. The nature and quantity of waste products

Light-water reactors discharge variable amount of radionuclides into the atmosphere and waters, whether they be fission products such as the dangerous ^{131}I, ^{137}Cs or ^{90}Sr, or neutron-capture products (^{60}Co, ^{54}Mn or ^{3}H). Table 4.5 gives an estimate of the quantities of radionuclides released into the environment from a PWR-type nuclear power plant of 1050 MW of electricity. It can be seen that the station discharges 5620 Ci of ^{85}Kr into the air and 4000 Ci of tritium into the waters every year.

It should be noted, however, that the waste products vary considerably from one reactor to another according to the specifications of the treatment plant. From some reactors the amounts of waste released into the environment are theoretically far below the maximum permissible limits set down by the International Commission on Radiological Protection (ICRP/CIPR). In France, for instance, the public health authorities, in line with international standards, require that waste products after dilution in river water should not exceed 20 pCi/l for all pure α, β, $\beta\gamma$, and γ emitters together, and 1000 pCi/l for tritium, measured separately. It remains to be proved whether these standards can be respected when there are several PWR reactors running on the same river.

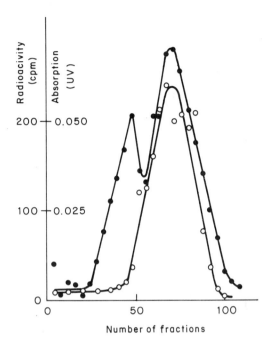

Fig. 4.10 The absorption of tritium taken by mouth into the DNA of the thymus gland of calves. Some animals who had been given predetermined amounts of tritium in their drinking water were slaughtered and their organs removed for analysis (particularly the testicles and thymus gland). the ultracentrifugation of an extract from a calf's thymus gland shows that the main peak of radioactivity (black dots) can be superimposed perfectly on the peak of maximum absorption into UV with the characteristic wavelength of nucleic acids (white dots). This proves the absorption of tritium into the DNA. (After Kirchmann et al., 1973. *Reproduced by permission of the International Atomic Energy Agency*)

The problem of tritium, although it should not be exaggerated as this element does not seem to accumulate in trophic chains, is still one to be reckoned with if light-water reactors increase in number. Even just a few PWR reactors constructed near the Mississippi at the end of the 1960s were enough to increase significantly the amount of tritium in the river water. In 1968, the level went up to 2000 pCi/l as against 10 pCi/l before the reactors were built (*in* Abrahamson, 1972). Furthermore, it has been shown that tritium can be absorbed into the nucleic acids with little difficulty when it is taken by mouth in water or contaminated food (Kirchmann *et al.*, 1973).

Another result of the frequent cracking of cladding is the release of radioiodine, estimated at 0.1 Ci per year on the site of a nuclear power station.

It is worth noting that it is now technically possible to build so-called 'zero release' nuclear power stations. A rise in the cost of the kilowatt hours caused by installing plants to recuperate liquid and gaseous effluents would not be prohibitive and may well dispel the justifiable caution that slows down the development of the PWR type of reactor due to their potential for water pollution.

Table 4.6 The annual waste dumped into the Channel (expressed in curies) from the spent nuclear fuel reprocessing plant at La Hague. (*In* Ancellin *et al.*, 1979)

Year	Tritium- 3H	Ruthenium-106 + Rhodium-106	Caesium-137	Strontium-90	Antimony-125	Plutonium	Total[1]
1966		389	197	28		0.03	642
1967		1326	443	11		0.32	1797
1968		1627	768	37		0.85	2453
1969		1434	545	19		0.36	2005
1970	1657	5409	2409	107	25	0.64	10 610
1971	2113	7712	6556	447	68	3.91	20 039
1972	2280	7571	890	868	492	1.79	13 867
1973	2967	7104	1872	510	1789	2.19	16 159
1974	7598	14 518	1913	2817	1862	15.7	32 878
1975	11 120	22 425	931	2029	1943	7.1	43 050
1976	7132	15 004	939	1078	971	4.2	26 430

[1] This number is greater than the sum of the preceding columns as we have only shown the principal radioelements discharged into the marine environment.

III — Pollution From Spent Nuclear Fuel Reprocessing Plants

1. Nature and importance

Whatever precautions may be taken with nuclear reactors, they do not solve the question of the waste produced during the spent fuel treatment phase. Even if this phase is not put under the same authority as that governing electricity production, the consequences of nuclear power processes cannot be ignored. The quantities of spent nuclear fuel are going to become considerable: they reach 27 tonnes per year for the PWR reactor of 1000 MW. It can be calculated that a programme for 50 GW of electricity based on light-water reactors will mean treating some 1550 tonnes per year of spent fuel (installed capacity from 1985 onwards in France). The French nuclear power programme will need, therefore, to treat about 5000 tonnes per year of highly radioactive spent fuel in the year 2000.

This means that all the fission products, neutron capture products and transuranium elements will be left over from the treatment processes designed to recuperate fissile matter which will take place in highly sophisticated treatment plants. These plants will be discharging into the atmosphere rare radioactive gases: ^{85}Kr, ^{133}Xe and low levels of various radioactive isotopes of iodine, especially ^{129}I and ^{131}I. Despite the many precautions taken, it is actually impossible to store all the radioiodine produced although luckily most of it is recovered. As for the noble gases, they are entirely discharged into the atmosphere. In addition, the treatment plants pour several thousand cubic metres per day of low level radioactive liquid effluents into the rivers and seas.

It would seem, therefore, that the most serious environmental problem caused by the growth of the nuclear industry is that of radioactive waste. As long as

Table 4.7 Growth projections on the production of radioactive waste by the nuclear power industry in the United States (After SCEP, 1970)

Annual production	1970	1980	2000
Installed nuclear power capacity (in MW)	11 000	102 000	1 090 000
Production of high activity waste (in gallons/year)[1]	23 000	548 000	5 050 000
Total volume produced (in gallons)	45 000	2 577 000	57 900 000
Total quantity of fission products (in megacuries):			
strontium-90	15	805	16 000
krypton-85	1.2	96.6	1722
tritium	0.04	3.2	53.4
Total for all fission products (in gallons)	1200	47 240	1 277 000
Total for all fission products (in tonnes)	16	416	7950

[1] US gallon = 3.785 litres.

there is no satisfactory technical solution, although this is no doubt being researched, one must agree that any significant development of nuclear power will be hindered.

2. The problem of radioactive waste

Treatment of considerable quantities of spent nuclear fuel will produce large amounts of waste.

Table 4.7 shows estimates made in the United States of the growth in nuclear power waste products between 1970 and the year 2000. It can be seen that by the end of this century the USA will have to deal with an accumulated quantity of waste equal to that produced by the explosion of several million Hiroshima-type bombs.

Managing such a mass of waste will have to be based on certain absolute principles: all biologically dangerous radioactive material must be isolated from the environment for more than 600 years, which is the length of time necessary for the decontamination factor to reach 10^6, in the case of elements with an average half-life of about 30 years. In fact, for a radioisotope's activity, and therefore its mass, to decrease 10^6 times it would mean waiting for 20 half-lives as $2^{-20} \simeq 10^{-6}$. For caesium-137, whose half-life is 32 years, it comes to 640 years. For plutonium, with a half-life of 24 000 years, it would mean disposing of waste contaminated with this element from the biosphere for 480 000 years.

As the amount of waste is proportional to the consumption of fissile matter, which is doubling every four years in France for instance, then no programme for disposing of low-level radioactive waste, especially of effluents diluted in the waters, should be undertaken unless it is certain that there will be no risk even when the amount of waste increases tenfold. There can be no compromise on this matter between safety and economic profitability.

The current solution that seems the safest is to store the waste in disused salt-mines. Although still controversial, it seems that the area of mines needed for the disposal of nuclear wastes is less than previously expected. Cohen (1977) has calculated that it would take less than 1 km² of mines per year to store the wastes from a nuclear power system with 400 GW installed capacity (about twice the present US needs for electricity).

IV — Contamination of Trophic Webs by the Radionuclides Discharged into the Environment by the Nuclear Industry

At the present time, atmospheric pollution caused by the nuclear industry is negligible, except perhaps for air pollution by ^{85}Kr around spent fuel

Table 4.8 Variation in the radioactivity (in pCi/kg of wet weight or per litre) of constituents of the Meuse River after the Chooz reactor was brought into service in April 1967. (After Micholet-Cote, *et al.*, 1973. *Reproduced by permission of the International Atomic Energy Agency*)

Constituents	Year of sampling	Radioelements		
		^{58}Co	^{137}Cs	^{90}Sr
Water	1966	0	1.7	0.19
	1971	2.3	4.46	0.86
Dry mud	1966	0	1700	140
	1971	1 000	60 000	92
Mosses	1966	0	60	250
	1971	500	9 600	—
Aquatic animals (except fish)	1966	—	100	900
	1971	120	1 000	—
Fish	1966	—	65	19
	1971	250	8200	74

Table 4.9 Report on the concentrations found in eviscerated fish from the Meuse River sampled in 1971. (Calculations based on the annual average concentrations in the water.) (After Micholet-Cote *et al.*, 1973. *Reproduced by permission of the International Atomic Energy Agency)*

Species	Radioactivity of the fish/of the waters of the *Meuse*					
	Radioelements					
	^{90}Sr	^{134}Cs	^{137}Cs	^{54}Mn	^{58}Co	^{60}Co
Gudgeon	100	178	90	647	n.d.	96
	127	645	471	863		138
Bream	170	411	291	144	n.d.	86
Rudd	400	411	336	144	n.d.	39
Pike	57	880	605	432	24	n.d.

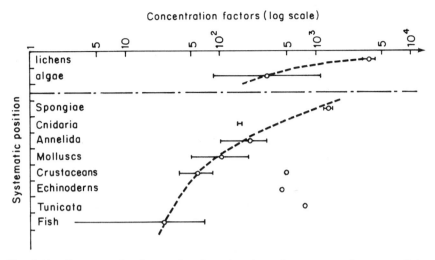

Fig. 4.11 Concentration factors for plutonium in various taxonomic groups: lichens, algae and marine animals. (After Fraizier and Guary *in* Abst. USERDA-AIEA conference on transuranians in the environment, San Francisco, November 1975, AIEA, Vienna, 1976, p 686)

reprocessing plants, which is, however, still below the maximum radiation levels allowed by the ICRP.

1. Contamination of limnic and marine ecosystems

Incorporation into the Trophic Chains

The pollution of continental waters and coastlines is becoming of great concern and there have already been cases of certain radionuclides accumulating in trophic webs.

(a) Contamination of limnic biocoenoses — In France, the construction of the Chooz nuclear power station on the Meuse River (1967) has resulted in significant contamination of the river's water, benthic silts and the biomass (cf. Table 4.8). Table 4.8 shows that in this well-defined case, installing a single reactor of low capacity (260 MW) was enough to increase considerably the radioactivity added to that of the surface waters (38 pCi/litre, excluding tritium, as opposed to the maximum of 20 pCi/litre allowed by the ICRP).

Still more worrying was the fact that processes of bioconcentration from this contamination caused an accumulation of various isotopes in the fish at a level of more than 24 to 880 times above the amounts found in the river water (cf. Table 4.9).

This example and many others show that the apparent safety afforded by the standards for passive dilution still leaves many doubts regarding the possibilities for concentration in the biomass.

(b) Contamination of marine biocoenoses—Processes of concentration of radionuclides in the biomass are of particular concern in the marine environment. Certain oceanic organisms seem to have a considerable capacity for reconcentrating radioelements dissolved in the water. Also, the food chains are very long and food webs are very complex in these ecosystems.

Table 4.10 gives the estimated average concentration factors for the major categories of marine organisms. It shows that the concentration generally increases when passing from one trophic level to the next higher. However, in the case of radioelements with a high atomic mass, such as radium and plutonium, the capacity for bioconcentration by lower organisms is clearly greater than for species further up the trophic pyramid (fish, for example), (cf. Fig. 4.11).

The same is true for certain radionuclides of biogenic elements and for numerous other radioelements whose chemical properties are similar to those of some vital elements for metabolism. They too have a concentration factor (radioactivity of the organism/radioactivity of the sea-water) which decreases the further one goes up the systematic scale from the algae to the vertebrates.

Amiard (1978) studied the kinetics of the accumulation in experimental trophic chains of a few important radionuclides released in the effluents from a reprocessing plant into the marine environment: antimony-125, cobalt-60, strontium-90 and silver-110m. He defined the ways in which these radionuclides are transferred via water or food to an organism on a given level as well as their movement towards the higher trophic level. For his experiments he used several types of benthic food chains, both short (annelid, crustacean or annelid, fish) and long:

diatom →	mollusc →	crustacean →	mammal
(*Navicula ramosissima*)	(*Scrobicularia plana*)	(*Carcinus moenas*)	(*Rattus rattus*)
Trophic level I	II	III	IV

Amiard (1978) and Amiard-Triquet and Amiard (1976) concluded from their own experiments and research work in foreign laboratories that two types of contamination by radionuclides of marine food webs must be distinguished. The first is by elements that tend to build up (such as ^{137}Cs and certain transition metals) with their concentrations increasing significantly as they go up the ecological pyramid. The second type of contamination, and the most general, is by quite a considerable number of elements, such as ^{60}Co, ^{135}Sb, ^{110m}Ag, which simply transfer from one level to another in the trophic chain without any bioamplification occurring.

Amiard (1978) observed that with cobalt-60 there were relatively significant fluctuations in the radioactivity of ingested food within his long food chain, without the radioactivity of the experimentally contaminated organisms being affected. This means that there is no direct correlation between the level of ^{60}Co in a marine invertebrate and the amount contained in its food. This phenomenon can be explained by a process of regulation of the element within the animal organism which only retains the necessary quantity for its metabolism, not absorbing the rest of the element or, at most, storing it in the hepatopancreas, for example, before excreting it.

It should be noted, however, that even in cases where there is no bioconcentration of an element at the upper trophic level, with the transfer factor (TF) between two successive trophic levels remaining less than one, the concentration factor (CF) of the given organism from sea-water, and, therefore, its global contamination (C), still remain more than one (cf. Fig. 4.12).

It should also be noted that the physiological regulation of absorption of a radioactive element can still have damaging radiation effects because the levels of concentration in the storage organ (such as the liver) amount to serious internal irradiation. Furthermore, even if the global concentration decreases along a food chain, the radiation doses to which the organisms are exposed can rise for species situated at the end of the chain, as size (and therefore mass) of organisms generally increases as one goes up the ecological pyramid. On the other hand, for biogenic elements such as ^{32}P, ^{14}C or ^{55}Fe, the concentration found in molluscs and fish is comparable and even greater than that found in phytoplankton.

It is worth pointing out at the bottom of Table 4.10 that, contrary to often widespread opinion, plutonium is capable of being concentrated in the biomass with an average concentration factor of 200 times for molluscs which goes up to as much as 1300 times for plankton. For instance, the exposure for 15 days of a cephalopod, *Octopus vulgaris*, to sea-water contaminated with 30 nCi/l of plutonium-237 shows that the mollusc concentrates this element in its organism. The average concentration factor is 40 for the whole animal but reaches 9300 in their branchial hearts compared to the sea-water. It has, therefore, been suggested to use this cephalopod as a bioindicator of ocean pollution from transuranium elements (Guary *et al.*, 1981). In another study, it was found that americium-241 accumulates especially in the adrenochromes (pigmentary concretions) of their branchial hearts (Miramand and Guary, 1981). Recently, it has also been shown that terrestrial plants are able to concentrate plutonium through their roots.

A Few Examples of the Underestimation of Nuclear Pollution

Miscalculation of these processes of bioconcentration have given rise in the past to a few serious errors regarding the radioprotection of human populations.

For instance, on the Columbia River, which is polluted by the Hanford plutonium reactors, fish consumers have been exposed to doses of ^{32}P 40 000 times higher than levels in the river water, which means they ingested 40 nCi of ^{32}P per kilogram of fish consumed per year. The fishermen and other fish-eaters were, therefore, being exposed to 0.3 rem per year, nearly double the permitted maximum for human populations (170 mrem). Also in the Columbia

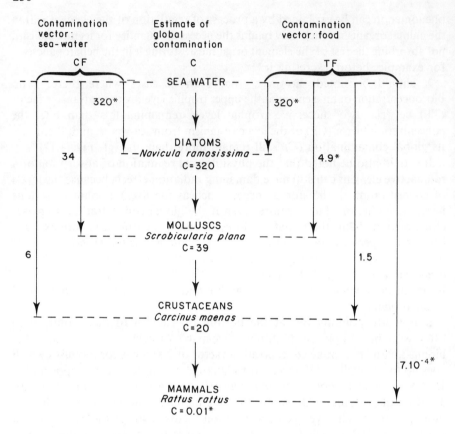

Fig. 4.12 The accumulation of cobalt 60 in organisms of an experimental marine food chain as a function of the various vectors of contamination CF, C and TF are the respective concentration factors. (After Amiard and Amiard-Triquet, 1976)

River ecosystem, the concentration of ^{32}P for the birds situated at the end of the food chain was 600 000 compared to the river water.

Another classic example of contamination of human food by the nuclear industry is that of Windscale in Great Britain: this nuclear fuel reprocessing plant was discharging 2000 Ci per month of ^{106}Ru into the Blackwater estuary in Cumbria. The sea-water was contaminated by 80 pCi/l. Tens of thousands of English people were eating the famous 'laverbread' made from the red alga *Porphyra umbilicalis* which concentrates ^{106}Ru at a level 1800 times greater than that found in sea-water. Certain people who were consuming large amounts of laverbread were exposed to 1.6 rem in their rectum. Even now, in spite of all the measures taken over the past decade or more, English people who consume more than 120 g/day of laverbread are being exposed to 5 times the maximum radiation dose allowed by the ICRP/CIPR for human populations (170 mrem/year).

Table 4.10 The average concentration factor for various radioisotopes of biological importance in the marine environment. (After Rice and Wolfe in Hood, 1971)

Radionuclides	Algae	Crustaceans	Molluscs	Fish
^{3}H	0.90	0.97	0.95	0.97
^{7}Be	250	—	—	—
^{14}C	4000	3600	4700	5400
^{24}Na	1	0.2	0.3	0.13
^{32}P	10^4	2×10^4	6×10^4	3.7×10^4
^{45}Ca	2	120	0.4	1.2
^{45}Sc	1200	300	—	750
^{51}Cr	2000	100	400	100
$^{54, 56}$Mn	3000	2000	10^4	1000
$^{55, 59}$Fe	2×10^4	2500	10^4	1500
$^{57, 58, 60}$Co	500	500	500	80
^{65}Zn	10^3	2000	1.5×10^4	1000
^{85}Kr	~1	~1	~1	~1
$^{89, 90}$Sr	50	2	1	0.2
$^{90, 91}$Y	500	100	15	10
^{95}Zr, ^{95}Nb	1500	100	5	1
^{103}Ru, ^{106}Ru, ^{106}Rh	400	100	10	1
^{110}Ag	—	7	10^4	—
^{132}Te, ^{132}I	—	—	—	—
^{131}I	5000	30	50	10
^{133}Xe	~1	~1	~1	~1
^{137}Cs	15	20	10	10
^{140}Ba, ^{140}La	25	—	—	8
$^{141, 144}$Ce	700	20	400	3
$^{185, 187}$W	5	2	20	3
$^{203, 210}$Pb	700	—	200	—
^{210}Po	1000	—	—	—
^{226}Ra	1000	100	1000	—
^{239}Pu	1300	3	200	5

2. Contamination of agroecosystems

Absorption by Plants

Another worrying aspect of diluted liquid waste is connected with the use of continental waters for crop irrigation.

In France, for example, agronomists are particularly concerned about the fact that the waters of the Rhône are used for irrigation when pollution levels in the river are continually growing, despite all the precautions taken, due to existing or projected nuclear power plants throughout its length (Bugey, Creys-Malville, Pierrelatte, Tricastin, Marcoule, etc.).

Tables 4.11 and 4.12 show that irrigation by sprinklers causes crop contamination naturally just as easily as under experimental conditions and also pollutes plants internally by penetrating through their leaves.

The study of fruit and market-garden crops in the Rhône valley shows that their contamination has nothing to do with their distance from the nuclear power

Table 4.11 Crop contamination by Cesium 137 from irrigation by sprinkling in the Rhône valley. (After Bovard et al.,1973. *Reproduced by permission of the International Atomic Energy Agency*)

Distance from Marcoule to sampling point (in km)	Total activity expressed in pCi/kg fresh weight					
	Apples	Peaches	Grapes	Tomatoes	Lettuce	Spinach
0	1500	2100	1700	2600	—	3300
10	990	1800	1400	1900	8500	5200
20	1000	2400	1200	2100	3600	5600
50	800	1250	—	3000	—	3200
76	1200	1700	1400	2100	6400	—

Table 4.12 Transfer rate of ^{60}Co into crops (direct contamination in the form of cobalt chloride) calculated experimentally (in pCi/kg fresh weight). (After Delmas et al., 1973. *Reproduced by permission of the International Atomic Energy Agency*)

	Demineralized water	River water
Lettuce	4.9	1.57
Beans	10	3.5

station (Marcoule). It does not result from transport by air of the radionuclides but from the use of the river water for irrigation which contains diluted effluents. Bovard et al., (1973) point out a fundamental factor: although the concentrations they found, especially of ^{137}Cs, are considerably less than the maximum permissible concentration (MPC) in water, the levels are much higher than those calculated on experimental models (Table 4.11).

Given the projected expansion of the nuclear industry, greatly increasing the waste produced, it would be easy to be pessimistic about man's eventual exposure to radiation through his food supplies.

Table 4.12 shows experimentally that crop contamination is not just superficial, due to spraying the aerial parts of the plants. The radionuclides also penetrate the edible parts and become concentrated compared to their level when diluted in water. Furthermore, Bovard et al. state that the estimated radioactivity of the Rhône water at the point the waste is discharged is a maximum of 7.3 pCi/l of ^{137}Cs. The contamination of 1 kg of polluted crops before being washed is, therefore, equal to that of 20 litres of irrigation water.

Absorption Through the Roots

Even more damaging to crops in the long term can be contamination through the roots from polluted soil. Due to their clay-humus colloidal complex, soils are capable of high levels of absorption of certain ions diluted in water, especially Sr^{2+} and above al Cs^+.

In theory, 1 gram of smectite, an important clay in the absorbing function in soils, can retain 6.3 Ci of ^{137}Cs (*in* Guenelon, 1970). The mineral colloids of the clay–humus colloidal complex can, therefore, fix various radioisotopes of biogenic elements to such an extent that a dangerous level of soil contamination can result, even from irrigation water containing a very low level of pollution. These radionuclides will then penetrate through the roots and be absorbed by the crops.

On the other hand, some ions such as ^{106}Ru are relatively mobile in soils and can spread horizontally for several kilometres from the source of pollution or contaminate deep ground waters.

3. The problem of external irradiation

Emission into the environment of certain radionuclides, even if they are chemically inert, can have ecological consequences. 'Isolated individuals' living near atomic power plants, workers in the nuclear industry and, more generally, all human populations can in fact be subjected to external irradiation.

This is true for some rare radioactive gases, such as ^{133}Xe and especially ^{85}Kr. Noble gases, because of their very lack of chemical reactivity, accumulate in the atmosphere, generally following their discharge from nuclear fuel reprocessing plants.

Table 4.13 summarizes the amounts of gases emitted and accompanying irradiation of populations living near a reprocessing plant capable of treating with the Purex process spent fuel from a nuclear power station with an installed capacity of 50 GW.

Table 4.13 A summary of the amounts of gases emitted and accompanying irradiation of populations living near a spent nuclear fuel reprocessing plant. (After Beaujean *et al.*, 1973. *Reproduced by permission of the International Atomic Energy Agency*)

Radionuclides	^3H	^{85}Kr	^{129}I	^{131}I
Emission in Ci/year	1.3×10^6	2.7×10^7	73	5.9×10^3
Total exposure in mrem/year	77	328	1.9×10^4	8.6×10^4
Necessary decontamination factor	20	500		2000

Taking into account the recommendations of the ICRP/CIPR, which advocates maximum exposure of 30 mrem/year for populations living near nuclear plants, it is clear that the quantity of radionuclides released will lead to higher radiation doses.

The equation:

$$D = D(^3H) + D(^{85}Kr) + 1/3 D(^{129}I + ^{131}I) = 30 \text{ mrem}$$

assumes quite high coefficients of decontamination.

Krypton 85

We should now consider what becomes of the ^{85}Kr which is being systematically released into the atmosphere of the northern hemisphere. It must be imagined evenly distributed through the troposphere between altitudes of 0 and 10 km, due to its long half-life (10.76 years) and chemical inertia and also the predicted expansion of the nuclear industry. Irradiation of human populations by this element, particularly affecting the lungs, will increase considerably in the future. Note that a concentration of 1 pCi/cm^3 of ^{85}Kr causes irradiation of the skin of about 2000 mrem/year, 20 mrem/year of the gonads and is estimated at between 14 and 23 mrem/year for the whole body (*in* Rohwer, 1973).

If the growth of the nuclear industry continues along present lines, by the year 2000 it would cause the exposure dose to go above the 30 mrem/year permitted maximum for irradiation from gaseous radionuclides. This will affect all human populations (Coleman and Liberace *in* Cook, 1971).

Furthermore, the excessive amounts of krypton-85 released into the atmosphere may well cause unfavourable climatic modifications by increasing the ionization of the atmosphere (Boeck, 1976). Current levels of ^{85}Kr discharged into the troposphere could totally change the electrostatic charge of the troposphere within the next half-century. Although there is no way of predicting whether these changes will be good or bad, especially for rainfall patterns, the global scale of this upheaval would require several decades for there to be a return to normal after ^{85}Kr emissions cease.

Consequently, it is urgent that from now on the rare radioactive gases from nuclear fuel reprocessing plants should be collected rather than released into the atmosphere without any thought for the long-term effects. There are several techniques now available for krypton and other rare gases, so the nuclear industry has no valid excuse for not using them.

The Case of People Working in Radioactive Zones

Workers in the nuclear industry are among the most exposed groups of individuals to nuclear pollution.

Nuclear industry personnel are particularly exposed to risks from radiation, some of them with the possibility of inhaling plutonium dust, etc. (At the American Kerr MacGee Company which extracts plutonium from spent fuel, there were 73 inhalation accidents among 100 workers in the period from April 1970 to August 1974.)

The accident in 1974 to the heat exchanger of the PWR reactor at Indian Point I (560 MW) in the United States provided valuable information on this subject (Smith, 1974). It needed 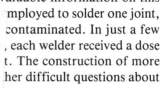 welders to be employed to solder one joint, as the primary circuit in the reactor was so highly contaminated. In just a few minutes of working on the defective heat exchanger, each welder received a dose of several rems from the reactor circuit. The construction of more reactors of this type in the future raises other difficult questions about radioprotection of the workers.

The risks of accidents in spent nuclear fuel reprocessing plants are no less significant, whenever such work is carried out. Enforced closure of the Nuclear Fuel Services plant in the United States by the controlling authorities (National Regulatory Commission; NRC), following inadequate radioprotection measures, is an example of how the problems of safety at work in the nuclear industry will only increase in the future, given the industry's present considerable growth.

Ultimately, we must emphasize on the severe irradiation hazards consecutive to a major accident occurring in a commercial nuclear reactor. As a consequence of the Tchernobyl disaster in April 1986, over 300 people inside the perimeter of the power plant were severely irradiated of whom 32 died subsequently.

Conclusion

Reading this book, it appears to have been limited to the most conventional concerns of ecotoxicology which are to study the ways that toxic substances contaminate the environment and to analyse the ecological consequences of their action. This would seem to contradict the definition of ecotoxicology given at the beginning of the book, namely that it includes the study of all the pollutants with which modern man contaminates the ecosphere.

In actual fact, constraints of size have meant that various important problems have had to be excluded here: organic pollution of water, eutrophication of lakes, thermal pollution and the interaction between pollutants and biogeochemical cycles (although the latter was mentioned in connection with sulphur dioxide).

A few general conclusions can now be drawn, in spite of the important omissions of which we are only too well aware.

Firstly, the study of ecotoxicology means making a clean sweep of a whole range of preconceived ideas and narrow administrative conceptions belonging to contemporary technocratic ideology. It quickly shows up the stupidity of notions of political frontiers and the futility of national legislations in the face of current burning issues (pollution of the oceans by petroleum, contamination of international waterways like the Rhine river, role of transboundaries transport of atmospheric pollutants from western Europe to Scandinavia, or from North eastern United States to east of Canada in the extension of the acid rains phenomenon for example).

Ecotoxicology also teaches how hazardous are the standards of 'inert' dilution set by the authorities concerned: bioamplification phenomena too often 'omitted', have already called into question the validity of such criteria (in the case of mercury or pollution of water by radionuclides).

Thirdly, ecotoxicology, through its numerous experimental contributions, questions the whole principle of manufacture and large-scale commercialization of non-biodegradable substances.

From now on, all of the industrialized countries should harmonize their legislative arsenal relating to licensing of chemical products. Only synthetic biodegradable substances of which the products of their biodegradation are harmless by themselves should be allowed for widespread use. Tests of the ecological impact of a new product should be made obligatory to ensure that before it is licensed it is not likely to accumulate in trophic chains.

The accident in 1974 to the heat exchanger of the PWR reactor at Indian Point I (560 MW) in the United States provided valuable information on this subject (Smith, 1974). It needed 2000 welders to be employed to solder one joint, as the primary circuit in the reactor was so highly contaminated. In just a few minutes of working in the defective heat exchanger, each welder received a dose of several rems from the high level of ^{60}Co present. The construction of more reactors of this type will no doubt in future raise other difficult questions about radioprotection of personnel.

The risks of accidental contamination in spent fuel reprocessing plants are no less significant, especially when maintenance work is carried out. Enforced closure of the Nuclear Fuel Service (NFS) plants in the United States by the controlling authorities (National Regulatory Commission; NRC), following inadequate radioprotection measures, is an example of how the problems of safety at work in the nuclear industry will only increase in the future, given the industry's present considerable growth.

Ultimately, we must emphasize on the severe irradiation hazards consecutive to a major accident occurring in a commercial nuclear reactor. As a consequence of the Tchernobyl disaster in April 1986, over 300 people inside the perimeter of the power plant were severely irradiated of whom 32 died subsequently.

Conclusion

Reading this book, it appears to have been limited to the most conventional concerns of ecotoxicology which are to study the ways that toxic substances contaminate the environment and to analyse the ecological consequences of their action. This would seem to contradict the definition of ecotoxicology given at the beginning of the book, namely that it includes the study of all the pollutants with which modern man contaminates the ecosphere.

In actual fact, constraints of size have meant that various important problems have had to be excluded here: organic pollution of water, eutrophication of lakes, thermal pollution and the interaction between pollutants and biogeochemical cycles (although the latter was mentioned in connection with sulphur dioxide).

A few general conclusions can now be drawn, in spite of the important omissions of which we are only too well aware.

Firstly, the study of ecotoxicology means making a clean sweep of a whole range of preconceived ideas and narrow administrative conceptions belonging to contemporary technocratic ideology. It quickly shows up the stupidity of notions of political frontiers and the futility of national legislations in the face of current burning issues (pollution of the oceans by petroleum, contamination of international waterways like the Rhine river, role of transboundaries transport of atmospheric pollutants from western Europe to Scandinavia, or from North eastern United States to east of Canada in the extension of the acid rains phenomenon for example).

Ecotoxicology also teaches how hazardous are the standards of 'inert' dilution set by the authorities concerned: bioamplification phenomena too often 'omitted', have already called into question the validity of such criteria (in the case of mercury or pollution of water by radionuclides).

Thirdly, ecotoxicology, through its numerous experimental contributions, questions the whole principle of manufacture and large-scale commercialization of non-biodegradable substances.

From now on, all of the industrialized countries should harmonize their legislative arsenal relating to licensing of chemical products. Only synthetic biodegradable substances of which the products of their biodegradation are harmless by themselves should be allowed for widespread use. Tests of the ecological impact of a new product should be made obligatory to ensure that before it is licensed it is not likely to accumulate in trophic chains.

Finally, these technical measures must be accompanied by a radical change, or even 'mutation' in the way of thinking of the 'decision-makers'. It is high time that the voice of biologists, and ecologists especially, is heard by government authorities and those responsible for industry if we want to avert this plague of modern times, environmental pollution.

Bibliography

GENERAL WORKS

Amiard-Triquet, C. and Amiard, J. C. *Radioécologie des milieux aquatiques.* Masson, 1980, 200 pp.
Ancellin, J., Guegueniat, P., and Germain, P. *Radioécologie marine.* Eyrolles, 1979, 256 p.
Bellan, G. and Peres, J. M. *La pollution des mers.* Coll. 'Que sais-je?', P.U.F., 1974, 128 pp.
O'Brien, R. D. *Insecticides: action and metabolism.* New York, Academic Press, 1967, 332 pp.
Butler, G. C. *Principles of ecotoxicology.* Wiley and Sons, 1978, 350 pp.
Chadwick, M. J. and Goodman, G. T. *The ecology of resources degradation and renewal* 15[e] symp. of the British ecological society. Oxford, Blackwell Scientific Publications, 1975, 480 pp.
Ferry, B. W., Baddeley, M. S., and Hawksworth, D. L. *Air pollution and lichens.* The Athlone Press, London University, 1973, 390 pp.
Gerlach, S. A. *Marine pollution, diagnosis and therapy.* Springer-Verlag, Berlin, 1981, 218 pp.
Gervais, P. *Allergologie et ecologie.* Masson, Paris, 1976, 150 pp.
Hartung, R. and Dinman, B. *Environmental Mercury contamination* Ann. Arbor Science Pub., 1972, 349 pp.
Holdgate, M. M. *A perspective of environmental pollution.* Cambridge University Press, 1979, 278 pp.
Hood, D. W. *Impingement of man on the ocean.* Wiley Interscience, 1971, 738 pp.
Lacaze, J. C. *La pollution pétrolière en milieu marin.* Masson, 1980, 132 pp.
Lamotte, M. and Bouliere, F. *Problèmes d'Écologie.* I: L'échantillonnage des peuplements animaux des milieux terrestres, Masson, 1969, 303 pp.
Lamotte, M. and Bouliere, F. *Problèmes d'Écologie.* II: L'échantillonnage des peuplements animaux des milieux aquatiques, Masson, 1971, 294 pp.
Lenihan, J. and Fletcher, W. W. *The chemical environment* in Environment and man, vol. 6, Blackie ed., 1977, 163 pp.
Matthews, W. H., Smith, F. E., and Goldberg, E. D. *Man's impact on terrestrial and oceanic ecosystems.* Cambridge Mass, MIT Press, 1971, 540 pp.
MacCormac, B. M. *Introduction to the scientific study of atmospheric pollution.* Dordrecht, Reidel, 1971, 169 pp.
Mellanby, K. *Pesticides and pollution.* London, Collins, 1971, 2nd ed, 222 pp.
Moore, J. W. and Ramamoorthy, R. *Heavy metals in natural waters.* New York, Springer Verlag, 1984, 268 pp.

Moore, J. W. and Ramamoorthy, R. *Organic chemicals in natural waters*. New York, Springer Verlag, 1984, 290 p.
Moriarty, F. *Ecotoxicology, the study of pollutants in ecosystems*. London, Academic Press, 1983, 73 pp.
Muirhead-Thomson, R. C. *Pesticides and freshwater fauna*. London and New York, Academic Press, 1971, 248 pp.
Odum, E. P. *Fundamentals of ecology*. Philadelphia and London, Saunders, 1971, 2nd ed., 573 pp.
Peres, J. M. et al. *La pollution des eaux marines*, Gauthiers-Villars, 1976, 240 pp.
Pesson, P. et al. *La pollution des eaux continentales*, Gauthier-Villars, 1976, 285 pp.
Polikarpov, G. G. *Radioecology of aquatic organisms*. North-Holland Publishing Company, 1966, 314 pp.
Ramade, F. *Éléments d'écologie appliquée*. Paris, MacGraw-Hill, 1978, 2nd ed., 576 pp.
Ramade, F. *Ecology of natural resources*. Chichester and New York, John Wiley, 1984, 232 pp.
Sacchi, C. F. and Testard, P. *Écologie animale*, Doin, 1971, 480 pp.
Simmons, I. G. *Ecology of natural resources*. London, Edward Arnold, 1974, 424 pp.
Singer, S. F. *Global effects of environmental pollution*. Dordrecht, Holland, Reidel, 1970, 218 pp.
Smith, J. E. *'Torrey-Canyon': pollution and marine life*. London, Cambridge U.P., 1968, 210 pp.
Truhaut, R. *Écotoxicologie et protection de l'environnement*. Abst. Col 'Biologie et devenir de l'homme', MacGraw-Hill—Édiscience, 1976, pp. 101–121.
Watt, K. F. *Principles of environmental sciences*. MacGraw-Hill, 1973, 320 pp.

ORIGINAL PUBLICATIONS AND REVIEWS

Abrahamson, D. E. Ecological hazards from nuclear power plants. In Farvar and Milton, *'Careless technology'*, 1972, Natural History Press.
Adamoli, P. et al. Analysis of 2, 3, 7, 8 tetrachlorodibenzo-*p*-dioxin in Seveso area. In Ramel, *Chlorophenoxyacetic acids and their dioxins. Ecol. Bull. Stockholm, 1978,* **27**, 31–38.
Advisory Committee on the Biological Effect of Ionizing Radiation. *Rep. on the effects on populations of exposure to low levels of ionising radiation*. Washington, DC 20006, Nat. Acad. Science U.S., Nov. 1972, 217 pp.
Akesson, B. The use of certain Polychaetes in bioassay study. *Rapp. Reun. Cons. Int. Explor. Mer*, 1980, **179**, 315–21.
Albers, M. H. Effects of external application of fuel oil on hatchability of mallard eggs. In Douglas and Wolfe, ed. *Fate and effects of petroleum hydrocarbons in marine ecosystems and organisms*. Pergamon press, 1977, pp. 158–163.
Albers, P. H. The effects of petroleum on different stages of incubation in birds eggs. *Bull. Env. Toxicol.*, 1978, **19**, 624–30.
Albers, P. H. and Szaro, R. C. Effects of n°2 fuel oil on common eider egg. *Marine Pollution Bull.* 1978, **9**, 138–139.
Almer, B. Försurningens inverkau pa fiskbersstand i västkustsjöar (effets de l'acidification des eaux sur les poissons). *Inf. Sötvattensbalrratonet Drottningholm*, no. 12, 1972, 47 pp.
Almer, B., Dickson, W., Ekstrom, C. et al. *Effects of acidification on Swedish lakes Ambio*, 1974, **3**, pp. 30–36.
Ames, B. N., Lee, F. D., and Durston, W. E. An improved bacterial test system for detection and classification of mutagens and carcinogens. *Proc. Nat. Acad. Sci (USA)*, 1973, **70**, 782–86.

Ames, B. N., MacCann, J., and Yamasaki, E. Carcinogens are mutagens: a simple test system. *Mutation Res.* 1975, **33**, 27-28.
Ames, B. N. *et al.*, and Sarasin, A. Des tests simples pour détecter les cancérogènes. *La Recherche*, Nov. 1975, **61**, 974-976.
Amiard, J. C. Contribution à l'étude de l'accumulation et de la toxicité de quelques polluants stables et radioactifs chez des organismes marins. Thèse de Doctorat es Sciences, Université de Paris-VI, 4 mars 1978, 147 p., 20 fig.
Amiard-Triquet, C. and Amiard, J. C. La pollution radioactive au milieu aquatique et ses conséquences écologiques. *Bull. écol.*, 1976, **7**, 3-32.
Anderson, D. W., Hickey, J. J., Risebrough, W. L. *et al.* Eggshell changes in certain North American birds. *Leiden, Proc. XVth Int. Cong. Ornithology*, 1972, 514-40.
Anderson, W. L. Lead poisoning in waterfowl at Rice lake, Illinois. *Journ. Wildl. Manage.*, 1975, **39** 264-70.
Arnoux, A. and Bellan-Santini, D. Relations entre la pollution du secteur de Cortiou par les détergents anioniques et les modifications des peuplements de *Cystoseira stricta*. *Tethys*, 1972, **4**, 583-6.
Ashton, P. S. Long term changes in dense and open inland forest following herbicide attack. In Westing, A. H. *Herbicides in war*. Stockholm International Peace Institute, Taylor and Francis, 1984, pp. 33-37.
Aubert, M., Petit, L., Donnier, B., and Barelli, M. Transfert de polluants métalliques au consommateur terrestre à partir du milieu marin. *Rev. Intern. Océanog. Méd.*, 1973, **30**, 39-59.
Bage, G., Cekanova, E., and Larsson, K. S. Teratogenic and embryotoxic effects of the herbicides di-and trichlorophenoxyacetic acids (2,4 D and 2,4,5 T). *Acta pharmacol. and toxicol.*, 1973, **32**, 408-16.
Baker, J. M. The impact of oil pollution on living resources. First draft of report by oil pollution working group. IUCN Commission on Ecology, Gland, Switzerland, 1982, 39 pp.
Barrows, H. L. Soil pollution and its influence on plant quality. *Journ. Soil Wat. Conserv.*, 1966, **21**, 211-6.
Batley, G. E. and Gardner D. A study of copper, lead and cadmium speciation in some estuarine and coastal marine waters. *Estuarine and Coastal Marine Science*, 1978, **7**, 59-70.
Beamish, R. J. Loss of fish populations from inexploited remote lakes of Ontario, Canada as consequence of atmospheric fallout of acid. *Wat Res.* 1974, **8**, 85-95.
Beaujean, M., Bohnenstingl, J., Laser, M. *et al.* Gaseous radioactive emissions from reprocessing plants and their possible reduction. *Environmental behaviour of radionucleides released by nuclear industry*, AIEA, Vienna, 1973, 63-78.
Beir Committee. The effects on populations of exposure to low levels of ionizing radiation. National Academy of Sciences, Washington, 1980, 638 pp.
Bellan, G. La pollution par les tensio-actifs, *in* Peres, *La Pollution des eaux marines*. Gauthier-Villars, Paris, 1975, pp. 31-50.
Bellan, G. and Peres, J. M. État général des pollutions sur les côtes méditerranéennes de France. *Quad. Civica Stazione. Idrobiol.* Milano, May 1970, **1**, 36-65.
Bellan, G., Reish, D. J., and Foret, J. P. Action toxique d'un détergent sur le cycle de développement de la polychète *Capitella capitata*. *C.R.Ac. Sci. Paris*, 1971, **272**, 2476-79.
Bellan, G., Foret, J. P., Foret-Montardo, P., and Kaim-Malka, R. H. Action *in vitro* de détergents sur quelques espèces marines. *Marine Pollution and Sea Life*, Déc. 1972, 1-4.
Berg, W., Johnels, A., Sjostrand, B., and Westermark, Y. Mercury content in feathers of swedish birds from the past 100 years. *Oikos*, 1966, **17**, 71-83.

Bleakley, R. J. and Boaden, P. J. S. Effects of an oil spill remover on beach meiofauna. *Ann. Inst. Océanog. Fr.* 1974, **50**, 51–8.

Bliss, C. I. The calculation of the dosage–mortality curve. *Ann. Appl. Biol.*, 1935, **22**, 134–65.

Blus, L. J., Andre, A. B., and Prouty, R. M. Relations of the brown pelican to certain environmental polluants. *Pest. Monit. Journ.*, 1974, **7**, 181–94.

Blus, L., Cromartie, E., Mac Nease, L., and Joanen, T. Brown pelican: population status, reproductive success and organochlorine residues in residues in Louisiana, 1971–76. *Bull. Environ. Contam. Toxicol.*, 1979, **22**, 128–35.

Blus, L. J., Neely, B. S., Belisle, A. A., and Prouty, R. M. Organochlorine residues in brown pelican eggs: relation to reproductive success. *Environ. Pollut.* 1974, **7**, 81–91.

Boeck, W. L. Meteorological consequences of atmospheric Krypton 85. *Science*, 1976, **193**, 195–98.

Bouche, M. B. Pesticides et lombriciens: problèmes méthodologiques et économiques. *Phytiatrie—Phytopharmacie*, 1974, **23**, 107–16.

Boudou, A. and Ribeyre, F. Comparative study of trophic transfer of two mercury compounds—$HgCl_2$ and CH_3HgCl—between *Chlorella vulgaris* and *Daphnia magna*. Influence of temperature. *Bull. Environ. Contam. Toxicol.*, 1981, **27**, 624–9.

Boulekbache, H., Levain, N., Puiseux-Dao, S., Ramade, F., Roffi, J., Roux, F., and Speiss, C. Effets cytopathologiques et physiotoxicologiques de certains insecticides, en particulier du lindane. Abstr. colloque com. europ. sur les problèmes inhérents à la contamination de l'homme et de son environnement par les pesticides persistants et les composés organohalogènes. Luxembourg, May 1974, 545–55.

Bovard, P., Grauby, A., Foulquier, L., and Picat,. P. Étude radioécologique du bassin rhodanien—strategie et bilan. Vienna, AIEA, 1973, 507–23.

Bowen, M. J. M. Natural cycles of the elements and their perturbation by man *in* Lenihan and Fletcher *The chemical environment: Environment and man*. Blackie, Glasgow and London, 1977, vol. 6, pp. 1–37.

Brar, S. S., Nelson, D. M. *et al.* Instrumental analysis of trace elements present in Chicago area surface air. *Journ. geophys. research*, 1970, **75**, 2939–45.

Breidenbach, A. W. Pesticides residues in air and water. *Arch. Environ. Health*, June 1965, **10**, 827–30.

Broecker, W. S., Bonebakker, E. R. and Rocco, G. G. The vertical distribution of cesium 137 and strontium 90 in the ocean. *Journ. Geophys. Res.*, 1966, **71**, 1999–2003.

Brown, C. C. The statistical analysis of dose–effect relationships *in* Butler G. C., ed. *Principles of ecotoxicology*. Wiley, 1978, pp. 115–48.

Brown, J. R. The effect of dursban on micro-flora in non saline waters. C. R. Coll. dispersion et dynamique des polluants dans l'environnement, in *Abstr. 3ᵉcong. int. chim. pest. Helsinki*, 1974, p. 506.

Bryan, G. W. Some aspects of heavy metal tolerance in aquatic organisms *in* Lockwood A. P., ed. *Effects of pollutants on aquatic organisms*. CUP, Cambridge, 1976, pp. 7–34.

Bryan, G. W. and Hummerstone, L. G. Adaptation of the polychaete *Nereis diversicolor* to estuarine sediments containing high concentrations of heavy metals. I—General observations and adaptation to copper. *J. Mar. Biol. Assoc. UK*, 1971, **51**, 845–63.

Bull, K. R., Murton, R. K., Osborn D. et coll. High level of cadmium in atlantic seabirds and sea-skaters. *Nature* 1977, **269**, 507–9.

Cairns, J. The cancer problem. *Scient. Amer.*, 1975, **223**, 64–78.

Cairns, J. (jr) Estimating hazard. *Bioscience*, 1980, **30** 101–7.

Carbiener, R. Etude écologique de la pollution par le mercure du bassin du Rhin. Rapport ver de Pharmacie, Université de Strasbourg, 1978, 154 pp.

Clark, D. R. Bats and environmental contaminants. Special Report-Wildlife n° 235, USDI, Fish and Wildlife Service, Washington D. C., 1981, 27 p.

Clark, D. R., Martin, C. O., and Swineford, D. M. Organochlorine insecticide residues in the free tailed bat. (*Tadurida brasiliensis*) at Brackencave, Texas. *Journ. Mammal.*, 1975, **56**, 429–43.

Clark, R. C. and MacLeod, W. Impacts, transports, mechanisms and observed concentrations of petroleum in the marine environment, *in* Malins, ed. *Effects of petroleum on arctic and subarctic marine environment organisms*. Academic Press, 1977, pp. 91–223.

Clausen, J., Braestup, L., and Berg, O. The content of polychlorinated hydrocarbons in arctic animals. *Abstr. 3^e cong. Int. chimie des pesticides, Helsinki*, 3–9/7/1974, Symposium sur la dispersion des pesticides dans l'environnement, p. 496.

Cohen, B. L. The disposal of radioactive wastes from fission reactors. *Scient. Amer.*, June 1977 p. 21–31.

Comar, C. L., Frum, B. F., Kuhn, U. S. G. *et al.* Thyroid radioactivity after nuclear weapons tests. *Science*, 1957, **126**, 16–18.

Comerzan, O. Bilan du cadmium dans l'environnement. Rapport du Ministère de l'environnement, comité des chaînes biologiques, Paris, 1976, ref 6041/1975, 120 pp.

Commoner, B. Threats to the integrity of the nitrogen cycle. in Singer *Global Effects of Environmental Pollution*, Reidel, 1970, pp. 70–95.

Conney, A. H., Miller, E. C., and Miller, J. A. The metabolism of methylated aminoazo dyes. V. Evidence for induction of enzyme synthesis in the rat by 3-methylcholanthiene. *Canc. Res.*, 1956, **T15**, 450–9.

Cook, E. Ionizing radiations *in* Murdoch: *Environment, Resources Pollution*, Sinauer, 1971, p. 267.

Copeland, R. A. Mercury in the Lake Michigan environment *in* Hartung and Dinman *Environmental mercury contamination*, Ann Arbor, 1972, pp. 71–92.

Cork, J. M. Gamma radiation and longevity of the flour beetle. *Radiation Research*, 1957, **7**, 551–7.

Coughtrey, P. J., Martin, M. M., and Youg, E. W. The woodlouse, *Oniscus asellus*, as a monitor of environment Cadmium levels. *Chemosphere*, 1977, **6**, 827–32.

Crosby, D. G. and Wong, A. S. Environmental degradation of 2,3,7,8-tetrachlorodibenzo-p-dioxin (T.C.D.D.). *Science*, 1977, **195**, 1337–38.

Curley, A. *et al.* Organic mercury poisoning identified as the cause of poisoning in human and hogs. *Science*, 1971, **172**, 65–7.

Davant, P., Fleury, A., and Petit, P. Le parc ornithologique du Teich. *Bull., Soc. Écol.*, 1975. **6**, 299–305.

David, D. and Lutz-Ostertag, Y. Etude comparative de l'action du DDT sur la mortalilé, la morphologie extreme et interne chez l'embryon de poulet et de caille. *C.R. AC. Sci. Paris*, 1972, **275**, 2171–3.

David, D. and Lutz-Ostertag, Y. Action de gonades d'embryons de poulets traités au DDT sur des embryons hôtes en greffc chorio-allantoidiennes. *C.R. Ac. Sci. Paris*, 1975, **280**, 1011–14.

Davis, H. C. and Hidu H. Effects of pesticides on embryonic development of clams and oysters on survival and growth of the larvae. *US Fish. Wildl Serv. Fisher Bull.*, 1969, **67**, p. 393–404.

De Bach *et al.*, in Huffaker. *Biological control*. Plenum Press, 1971, pp. 189–90.

Delmas, J., Grauby, A., and Disdier, R. Étude des transferts dans les cultures de quelques radionucleides présents dans les effluents des centrales électronucléaires. in *Environmental behaviour of radionucleides released by nuclear industry*, Vienna AIEA, 1973, 321–32.

De Witt, J. B. Effects of chlorinated hydrocarbon insecticides upon quail and pheasants. *J. Agric. Food Chem.*, 1955, **3**, 672.

Didier, R. Action du 2,4,5-T et de la Simazine sur les gonades de l'embryon de poulet et de caille en culture '*in vitro*'. *Bull. Soc. Zool., Fr.,* 1974, **99**, 93–99.

Didier, R. Étude du peuplement gonocytaire des ébauches gonadiques de l'embryon de caille après action de l'acide 2,4,5-T phénoxyacétique. *C. R. Soc. Biol.*, 1975, **169**, 574–80.
Didier, R. Etude de l'action d'un herbicide, le 2,4,5-T sur l'espèce avienne *Coturnix C. Japonica*. Thèse doctorat, Université de Clermont, 1981, 163 pp.
Dieter, M. P. Blood delta-aminolevulinic acid deshydratase (ALAD) to monitor lead contamination in canvas black ducks (*Aythya valisineria*). *Animals as monitors of environmental pollutants*. Nat. Acad. Sci., Washington, 1979, pp. 177–91.
Doubilet, C. Recherche de composés organohalogénés persistants dans le lait maternel. Thèse de Doctorat en médecine soutenue le 23 Juin 1981, Université de Paris VII, Faculté de médecine Bichat, 37 p. dactylo.
Druckrey *et al*. In Szakvary, A. *Les résidus de produits phytosanitaires et leurs incidences possibles*. Centre de recherches Foch Publ., 4 rue de l'Observatoire, 75005 Paris, 1965, 187 pp.
Duke, T. W., Lowe, J. I. and Wilson, A. J. Polychlorinated biphenyls in the water sediments and biota of Escambia Bay, Florida. *Bull. Envir. Contam. Toxicol.* 1970, **5**, 171–180.
Dupas, A. Les chlorofluorocarbones produits par l'homme sont-ils une menace pour l'ozone stratosphérique. *La Recherche*, Novembre 1975, **6**, 970–3.
Dustman, E. H., Stickel, L. F., Blus, L. J. *et al*. The occurrence and significance of polychlorinated biphenyls in the environment. *Trans. 36th conf. N. Am. Wild Nat. Res.*, 7–10 mars 1971, Wildlife Monog. Inst, ed. Wash. DC 20005, pp. 118–33.
Dustman, E. H., Stickel, L. F., and Elder, J. B. Mercury in wild animals lake St Clair, 1970. In *Environmental Mercury Contamination* Hartung and Dinman, Ann. Arbor Science Publishers, 1972, p. 46–52.
Duvigneaud, P. *La synthèse écologique*. Doin publishers, Paris, 1974, p. 81.
Epstein, S. The price we pay for 'progress': the hazards of the halogenated hydrocarbons *in* proc. int. conf. *Chemistry, environment and man*, Gottlieb Duttweiler Institute, Zurich, 1981, pp. 125–142.
Erne, K. Weed-killers and wildlife. *Proc. XIth Int. cong. game. biol. Stockolm*, 1973, SNV 1974, 13 E, pp. 415–22.
Ernst, W. Effects of pesticides and related compounds in the sea. *Helgol Meeresunters (Germany)*, 1980, **33**, 301–12.
Ferguson, D. M., Etherton, J. E., and Hayes, M. J. Mitochondrial increase after long-term feeding of morphamquat. *Nature*, 1969, **224**, 83–4.
Finley, R. B. Adverse effects on birds of phosphamidon applied to a Montana forest. *Journ. Wildlife Manag.*, 1965, **29**, 580–91.
Fisher, N. S. Chlorinated hydrocarbons pollutants and photosynthesis of marine phytoplanctons: a reassessment. *Science*, 1975, **189**, 463–4.
Fisher, R. A. and Yates, F. *Statistical tables for agricultural, biological and medical research*. Oliver and Boyd, London, 6th ed., 146 pp.
Fitzhugh, O. G., Nelson, A. A., and Quaife, M. L. Chronic oral toxicity of aldrin and dieldrin in rats and dogs. *Food Cosmet. Toxicol.* 1964, **2**, 551–62.
Flickinger, E. L. Wildlife mortality at petroleum pits in Texas. *J. Wildl. Manag.* 1981, **45**, 560–4.
Forstner, U. and Wittman, G. T. W. *Metal pollution in the aquatic environment*. Springer Verlag, Berlin, 1981, 486 pp.
Foster, R. F. and Rostenbach, R. E. Distribution of radioisotopes in the Columbia river. *Journ. Am. Wat. Wks. Ass.*, 1954, **46**, 640–63.
Fournier, E. Toxicité humaine des pesticides. *Bull. Soc, Zool. Fr.*, 1974, **99**, 19–131.
Fraizier, A. and Guary, J. C. Recherche d'indicateurs biologiques appropriés au controle de la contamination du littoral par le Plutonium. Abst. Int. Symp. on Transuranian nuclides in the environment. USERDA-AIEA, San Francisco, 12–17 November 1975. AIEA Vienna, 1976, ISBN 92-0-020076-1, 679–689.
Friberg *et al*. *Cadmium in the environment*, CRC Press, 1974, pp. 133–4.

Gaddum, J. H. Reports on biological standards. III—Methods of biological assay depending on quantal response. *Med. Res. Council Spec. Rep.*, 1933, **183**.

Gardiner, J. Complexation of trace metals by ethylenediaminetetraacetic acid (EDTA) in natural waters. *Water Research*, 1976, **10**, 507-14.

Gellert, D. J., Heinrichs, W. L., and Swerdloff D. S. DDT homologues: Estrogen-like effects on the vagina, uterus and pituitary of the rat. *Endrocrinology*, 1972, **91**, 1095.

Genelly, R. E. and Rudd, R. L. Effects of DDT, toxaphène and dieldrin on pheasant reproduction. *Auk.*, 1956, **73**, 529-39.

Georghiou, G. P. and Mellon, R. B. Pesticide resistance in time and space. In *Pest resistance to pesticides*, Georghiou and Sato, ed. Plenum Press, New York, 1983, 1-46.

Glover, H. G. Acidic and ferruginous mine drainages. In *The Ecology of Resource Degradation and Renewal.* Blackwell Scientific Publications, Oxford, 1975, pp. 173-195.

Goldberg, D., Bowen, V. T., Farrington, J. W. *et al.* The Mussel Watch. *Environ. Conserv.*, 1978, **5**, 101-25.

Goldberg, E. D. The Mussel Watch? A first step in global marine monitoring. *Marine Poll. Bull.*, 1975, **6**.

Goldwater, L. J. Mercury in the environment. *Scient. Amer.*, May 1971, **224**, 15-21.

Goldwater, L. J. and Stopford, W. Mercury in the Chemical Environment. Lenihan, J., and Fletcher, W. ed., Blackie, London, 1977, pp. 38-63.

Gregory, M. R., Kirk, R. M., and Mabin, M. C. G. Pelagic tar, oil, plastics and other litter in surface waters in the New Zealand sector of the Southern ocean, and on Ross dependency shores. *New Zealand Antarctic Record*, 1984, **6**, 12-28.

Grimm, H. Einflüsse und Gefahren der Anhäufung toxischen chemikalien in der Wechselwirkung zwischen Landschaft, Tier und Mensch. *Arch. Naturforsch. Landschafts.*, 1965, **5**, 203-12.

Guary, J. C., Higgo, J. J. W., Cherry, R. D., and Heyraud, M. High concentrations of transuranics and natural radioactive elements in the branchial hearts of the cephalopod *Octopus vulgaris*. *Marine Ecology. prog. ser.*, 1981, **4**, 123-6.

Guenelon, P. Pollution nucléaire. *Cah. Ing. Agron.*, 1970, no. 247, 95-7.

Harding, L. W. and Phillips, J. H. Polychlorinated biphenyls (PCB) effects on marine phytoplankton photosynthesis and cell division. *Mar. Biol.*, 1978, **49**, 93-101.

Harley, J. H. Plutonium in the environment: a review. *Rad. Res.*, 1980, **21**, 83-104.

Hart, L. G., Shultice, R. W., and Fout, J. R. Stimulatory effects of chlordane on hepatic microsomal drug metabolism in the rat. *Toxicol. Appl. Pharmacol.*, 1963, **5**, 371-86.

Hartung, G. The role of food chains in environmental mercury contamination, *in* Hartung, and Dinman *Environmental mercury contamination.* Ann. Arbor, 1972, pp. 172-201.

Harvey, G. R., Milkas, M. P., Boven, V. T., and Steinhauer, W. G. Observations on the distribution of chlorinated hydrocarbons in atlantic ocean organisms. *J. Wat. Rés. U.S.A.*, 1974, **32**, 103-118.

Havas, P. and Huttunen, P. Some special features of the ecophysiological effects of air pollution on coniferous forestry during winter, *in* Hutchinson and Havas, eds. *Effects of acid precipitation on terrestiral ecosystems.* Plenum Press, 1980, pp. 123-31.

Hawksworth, D. L. Mapping studies In *Air Pollution and Lichens*, Ferry, Baddelay and Hawksworth, Athlone Press, 1973, pp. 38-76.

Hawksworth, D. L., Rose, F., and Coppins, B. J. changes in the lichens flora in England and Wales attributable to pollution of the air by sulphur dioxide. In *Air Pollution and Lichens*, Ferry, Baddelay and Hawksworth, Athlone Press, London University, 1973, pp. 330-67.

Hay, A. Toxic cloud over Seveso. *Nature*, 1976, **262**, 636-38.

Hay, A. Red faces (and hot tempers) on 2,4,5-T. *Nature*, 1980, **286**, 97.

Hayes, W. and Dale, W. Storage of insecticides in French people. *Nature*, 1963, **199**, 1189.
Heath, R. G. Nationwide residues of organochlorine pesticides in wings of mallards and Black ducks. *Pest. Monit.*, Sept. 1969, **3**, 115–23.
Heath, R. G. and Hill, S. A. Nationwide organochlorine and mercury residues in wings of adult mallards and black ducks during the 1969–70 hunting season. *Pest. Monit.*, 1974, **7**, 153–74.
Heath, R. G., Spann, J. W., and Kreitzer, J. F. Marqued DDE impairment of Mallard reproduction in controlled studies. *Nature*, 1969, **224**, 47–8.
Heim, R. *Champignons Toxiques et Hallucinogènes*. Paris. N. Boubée ed., 1963, p. 327.
Heinz, G. Effects of low dietary levels of methyl mercury on mallard reproduciton. *Bull. Environ. Contam. Toxicol.*, 1974, **11**, 386–92.
Heinz, G. Effects of methylmercury on approach and avoidance behaviour of mallard ducklings. *Bull. Environ. Toxicol. Contam.*, 1975, **13**, 554–64.
Heinz, G. H., Haseltine, S. D., Hall, R. J., and Krynitsky, A. J. Organochlorine and mercury residue in snakes from pilot and spider islands, Lake Michigan—1978. *Bull. Environ. Contam. Toxicol.*, 1980, **25**, 738–43.
Helder, T. Effects of 2, 3, 7, 8-tetrachlorodibenzo-*p*-dioxin (TCDD) on early life stage of the pike (*Esox lucius*). *Science. Total. Environ.*, 1980, **14**, 255–265.
Hess, W. N. (ed). The *Amoco Cadiz* oil spill a preliminary scientific report, NOAA/EPA special report, 1978, 281 pp. 66 pl.
Hickey, R. J. Air pollution, *in* Murdoch, Environment, resources, pollution and society. *Sinauer*, 1971, 189–212.
Hickey, J. J. and Anderson, D. W. Chlorinated hydrocarbons and eggshell changes in raptorial and fish eating birds. *Science*, 1968, **162**, 271–273.
Holden, A. V. Source of polychlorinated biphenyls in the marine environment. *Nature*, 1970, **228**, 1220–1.
Holden, A. V. Mercury in fish and shellfish. A review. *J. Food. Technol. G.B.*, 1973, **8**, 1–25.
Horn, M. H., Teal, J. M., and Backus, R. H. Petroleum lumps on the surface of sea. *Science*, 1970, **168**, 245–6.
Horstadius, S. De l'effet désastreux de l'enrobage des semences sur l'avifaune suédoise. *L'Oiseau. Rev. fr. d'Ornith.*, 1965, **35**, 71–5.
Hoskins, W. M. Bioassay in entomological research. *Zeits. F. Pflanz und Pflanzenschutz*, 1947, **64**, 7–10.
Hoskins, W. M. Use of the dosage-mortality curve in quantitative estimation of insecticide resistance. *Misc. Publ. Entom. Soc. Amer.*, 1960, **2**, 85–91.
Hovette, C. le saturisme des anatides de Camargue. *Alauda*, 1972, **40**, 1–17.
Huffaker, C. B. *Biological control*. London and New York, Plenum Press, 1971, 511 pp.
Hunding, C., and Lange, R. Ecotoxicology of aquatic plant communities, *in* Butler, ed. *Principles of ecotoxicology*. Wiley, 1978, pp. 239–55.
Hunt, E. G. and Bischoff, A. I. Inimical effects on wildlife of periodic DDD application to clear lake. *Calif. Fish. Game*, 1960, **46**, 91–106.
Hutchinson, T. C. and Havas, M. (ed.). Effects of acid precipitations on terrestrial ecosystems. Proc. Nat. Conf. on effects of acid precipitation on vegetation and soils. Toronto, 1978, Plenum Press 1980, 654 pp.
Hsiao, Y. and Patterson, C. C. Lead aerosol pollution in the high sierra overrides natural mechanisms which exclude lead from a food chain. *Science*, 1974, **184**, 989–994.
Inman, R. E., Ingersoll, R., and Levy, E. A. Soil: a natural sink for carbon monoxide. *Science*, 1971, **173**, 1229–31.
IRPTC/UNEP Instruction for the selection and presentation of data for the international register of potentially toxic chemicals. UNEP, Geneva, 1979, 386 pp.
Jaffe, L. S. The global balance of carbon monoxide, *in* Singer *Global effects of environmental pollution*. Reidel, Dordrecht, 1970, pp. 34–49.

Jefferies, D. J. The role of the thyroid in the production of sublethal effects by Organochorine insecticides and polychlorinated biphenyls, *in* Moriarty, *Organochlorine insecticides: persistent organic pollutants*. Academic Press, 1975, pp. 131-230.

Jefferies, D. J., French, M. C., and Osborne, B. E. The effects of pp'-DDT on the rate, amplitude, and weight of the heart of the pigeon and bengalese finch. *British Poultry Sci.*, 1971, **12**, 387-99.

Jenkins, C. Étude de l'imprégnation plombique du pigeon biset (*Columbia Livia*) vivant en milieu urbain. *Terre et Vie*, 1975, **29**, 465-80.

Jenne-Levain, N. Études des effets du lindane sur la croissance et le développement de quelques organismes unicellulaires. *Bull. Soc. Zool. Fr.*, 1974, **99**, 105-109

Jensen, S. and Jernolov, A. Biological methylation of mercury in aquatic organisms. *Nature*, 1969, **223**, 753-4.

Jensen, S., Johnels, A. G., Olson, M., and Otterlind O. DDT and PCB in marine animals from Swedish waters. *Nature (UK)* 1969, **224**, 247-50.

Jernolov, A. Conversion of mercury fallout. In *Chemical Fallout*, C. C. Thomas. 1969, chap. 4.

Jernolov, A. Factors in the transformation of mercury to methyl mercury, *in* Hartung and Dinman, Environmental mercury contamination. Ann Arbor Science Publisher, Ann Arbor, Michigan, 1972, pp. 167-72.

Johnels, A. G. and Westermark, T. Mercury contamination of the environment in Sweden. In *Chemical Fallout*, C. C. Thomas, 1969, chap. **10**, p. 221-39.

Johnels, A. G., Westermark, T. *et al.* Pike and some other aquatic organisms in Sweden as indicators of mercury contamination in the environment. *Oikos*, 1967, **18**, 323-33.

Jones, E. P. DDT stopped, suit dropped. *Science*, 1971, **173**, p. 38.

Johnsen, I. Regional and local effects of air pollution, mainly sulphur dioxide, on lichens and bryophytes, *in* Hutchinson and Havas, eds. *Effects of acid precipitation on terrestrial ecosystems*. Plenum Press, 1980, pp. 133-40.

Keenleyside, M. H. A. Effects of forest spraying with DDT in New Brunswick on food of young atlantic salmon. *Journ. Fish. Res. Board. of Canada*, 1967, **24**, 807-22.

Kempf, C. and Sitler, B. La pollution de la zoocœnose rhénane par le mercure et les produits organochlores. *La Terre et la Vie*, 1977, **31**, 661-68.

Kerswill, C. J. Studies on effects of forest spraying with insecticides, 1952-63 on fish, and aquatic invertebrates in New Brunswick streams; introduction and summary. *J. Fisher. Res. Bd. Canada*, 1967, **24**, 701-8.

Kirchmann, R., Remy, J., Charles, P. *et al.* Distribution et incorporation du Tritium dans les organes des ruminants. *Abstr. Coll. environment. behaviour of radionucléides released in the nuclear industry*, AIEA Vienna, 1973, p. 385-412.

Knabe, W. Air quality criteria and their importance for forests. *Mitt. Forst. Bundes. Versuchanst. Wien*, 1971, **92**, 129-150.

Knauf, W. Bestimmung des Toxizität von tensiden bei Wasserorganismen. *Tenside Detergents*, 1973, **5**, 251-5.

Kochupillai, N., Verma, I. C., Grewal, M. S., and Ramalingeswami, V. Down's syndrome and related abnormalities in an area of high black ground radiation in coastal Kerala. *Nature*, 1976, **262**, 60-61.

Koeman, J. H., Oskamp, A. A. G., and Veen, J. Insecticides as a factor in the mortality of the sandwich tern (*Sterna sandvicensis*). A preliminary communication. *Med. Rijks Fac. Landbow. Wet. Gent.*, 1967, **32**, 841-854.

Koeman, J. H., Horsmans, T., and Maas, H. L. Accumulation of diuron in fish. *Med. Rijks Landbow. Gent.*, 1969, **34**, 428-33.

Koeman, J. H., Ten Noever De Brauw, M. C. and De Vos, R. H. Chlorinated biphenyls in fish, mussels and birds from the river Rhine and the Netherland coastal area. *Nature*, 1969, **221**, 1126-1128.

Korshgen, L. Soil—Food chain pesticide wildlife relationship in Aldrin treated fields. *Journ. Wildlife, Manag.* 1970, **34**, 186-99.

Kraybill, H. F., Broadhurst, M. G., Buckley, J. L., Spaulding, J. E., Wessel, J. R., Fishbein, L., Stickel, L. F., and Davies, T. Polychlorinated biphenyls and the environment. Interdepart. Task force on PCB's, Washington DC, Mai 1972, 181 p. US dept of commerce 72-10419.

Kupfer, D. Influence of chlorinated hydrocarbons and organophosphate insecticides in metabolism of steroids. *Ann. N. Y. Av. Sci.*, 1969, **160**, 244-53.

Lacaze, J. C. Ecotoxicology of crude oil and the use of experimental marine ecosystem. *Marine pollution Bull.*, 1974, **5**, 153-6.

Lacaze, J. C. Étude de l'influence des produits pétroliers sur la production primaire de l'environnement marin. Thèse de doctorat ès Sciences, University of Paris VI, 12 October 1978.

Laundon, J. R. Urban lichen studies. In Ferry, Baddeley and Hawksworth, *Air Pollution and Lichens*, University of London, 1973.

Legator, M. S., Kelly, F. J., and Green, S. Mutagenic effects of Captan. *Ann. N. Y. Ac. Sci.*, 1969, **160**, 344-51.

Lehner, P. N., and Egbert, A. Dieldrin and eggshell thickness in ducks. *Nature*, 1969, **224**, 1218-9.

Leonard, A., Delpoux, M., Decat, G., and Leonard, E. D. Natural radioactivity in southwest France and its possible genetic consequences for mammals. *Radiation Research* 1979, **77**, 170-81.

Levin, D. L. *et al. Cancer rates and risks.* DEHW Pub. n° (NIH) 75-691 (2e edn), Washington D. C. G. P. O., 1974.

Liang, T. and Lichtenstein, E. P. Synergism of insecticides by herbicides. *Science*, 1974, **186**, 1128-30.

Likens, G. E. and Bormann, F. H. Acid rain: a serious regional environmental problem. *Science*, 1974, **184**, 1176-8.

Likens, G. E., Wright, R. F., Galloway J. N., and Thomas J. Acid rains. *Scient. Amer.* 1979, **241**, 39-47.

Lillie, R. J. Air pollution affecting the performance of domestic animals. *Agricultural Handbook*, USDA, Washington DC 20402, 1970, n° 380, 109 pages.

Linzon, S. N. Effects of air pollutants on vegetation, in MacCormac, *Introduction to the scientific study of atmosphere pollution*, 1971, pp. 130-151.

List, R. J., Machta, L. T. and Allen, J. S. Strontium 90 on earth surface in radioactive fallout from nuclear weapons tests, A. W. Klement, ed. USAEC, Div. Techn. Inform., Oak Ridge, Tenn., 1965, pp. 359-368.

Lloyd, R. Toxicity testing with aquatic organisms: a framework for hazard assessment and pollution control. *Rap. Réunion. Cours. Int. Explor. Mer*, 1980, **179**, 339-41.

London, J. and Kelley, J. Global trends in total atmospheric ozone. *Science*, 1974, **184**, 987-9.

Lubinska, A. Acid rains threaten Black Forest: Germany's forests are dying *WWF. News*, 1982, n° 12, p. 8.

Luckey, T. D. and Venugopal, B. Metal toxicity in Mammals. I—Physiology and chemical basis of metal toxicity. Plenum Press Ed., 1977, 237 pp.

Lundholm, B. Interaction between oceans and terrestrial ecosystems, *in* Singer, *Global effects of environmental contamination*, Reidel, Dordrecht, 1970, p. 195.

Lutz, H. and Lutz, Y. Pesticides, teratogenese et survie chez les oiseaux. *Arch. Anat. Hist. Embr. Norm. Exp.*, 1973, **56**, 65-78.

Lutz, Y. and Lutz, H. Action néfaste de l'herbicide 2,4-D sur le developpement embryonnaire et la fecondité du gibier à plume. *C.R. Ac. Sci., Paris*, 1970, **271**, 2418-21.

Lutz-Ostertag, Y. and Lutz, H. Note préliminaire sur les effets œstrogènes de l'Aldrine sur le tractus urogénital de l'embryon d'oiseau. *C.R. Ac. Sci.*, 1969, **269**, 484-6.

Lutz-Ostertag, Y. and Lutz, H. Sexualité et pesticides. *Ann. Biol.*, 1974, **13**, 173–85.

Luxin, W. *et al.* Health survey in high background radiation areas in China. *Science*, August 1980, **209**, 877–80.

MacCann, J., Choi, E., Yamasaki, E., and Ames, B. Detection of carcinogens as mutagens in the *Salmonella* microsome test: assay of 300 chemicals. *Proc. Nat. Acad. Sci. USA*, 1975, **72**, 5135.

MacGregor, J. S. Changes in the amount and proportions of DDT and its metabolites, DDE and DDD in the marine environment of southern California, 1949–1972. *Fish. Bull.*, 1974, **72**, 275–93.

MacIntyre, A. D. and Pearce, J. B., eds. Biological effects of marine pollution and the problems of monitoring. Proc. ICES workshop held in Beaufort, North-Carolina, 26 Feb–2 March 1979, *in* Rept. meetings ICES, vol. 179, August 1980, 350 pp.

Marano, F. *Synchronisation et cycle cellulaire du Dunaliella bioculata, utilisation du modèle expérimental pour l'étude du mode d'action de l'acroléine.* Thèse soutenue le 17 Janvier 1980 à l'Université de Paris VII, 202 pp.

Marchand, M., Conan, G., and d'Ozouville, L. Bilan écologique de l'*Amoco Cadiz. Rapp. Scient. technique CNEXO*, 1979, **40**.

Mariotti, A., Letolle, R., Blavoux, B., and Chassaing, B. Détermination par les teneurs naturelles en N_{15} de l'origine des nitrates: résultats préliminaires pour le bassin de Melarchez (Seine et Marne). *C.R. Ac. Sci.*, 1975, **280**, 423–26.

Marshall, J. S. The effects of continuous gamma radiation on the intrinsic rate of natural increase of *Daphnia pulex. Ecology*, 1962, **43**, 598–607.

Martin-Bouyer. Enquête épidémiologique sur les conséquences de l'intoxication de nourrissons par l'hexachlorophène lors de l'accident du talc 'Morhange'. *Lancet*, 1982, **1**, 91.

Masters, R. L. Air pollution—Human health effects. In MacCormac, *Introduction to the Scientific Study of Atmospheric Pollution*, 1971, pp. 97–130.

Mastromatteo, E. and Sutherland R. B. Mercury in humans in the Great Lakes region, *in* Hartung and Dinman, *Environmental mercury contamination.* Ann. Arbor, 1972, pp. 86–92.

Matthews, W. H., Smith, F. E. and Goldberg, E. D. *Man's impact on terrestrial and oceanic ecosystems.* MIT Press, Cambridge, Mass., 1974, 540 pp.

Meiniel, R. Contribution à l'étude expérimentale de l'action des insecticides organophosphorés sur l'embryon d'oiseau. Analyse de l'effet tératogène et de certains troubles physiologiques. Thèse doctorat ès-sciences Naturelles. Université de Clermont, 1975.

Menhinick, E. F. Comparison of the invertebrate populations of soil and litter of moved grasslands in areas treated and untreated with pesticides. *Ecology*, 1962, **43**, 556–61.

Menser, H. A. and Heggestad, H. E. *Science*, 1966, **153**, 424–5.

Micholet-Cote, C., Kirchmann, R. *et al.* Étude de la radiocontamination des poissons de la Meuse. In *Environmental behaviour of radionucleides released in the nuclear industry.* AIEA., Vienna, 1973, pp. 413–27.

Miettinen, J. K. The present situation and recent developments in the accumulation of Cs^{137}, Sr^{90} and Fe^{55} in arctic foodchains. *Agence Atom. Intern. de Vienne,* Proc. sem., 1969; *Environmental contamination by radioactive materials*, 1969, pp. 145–51.

Miller, D. R. Models for total transport *in* Butler, *Principles of ecotoxicology.* Published on behalf of Scope by Wiley, 1978, pp. 71–90.

Milstein, R. and Rawland, F. S. Quantum yield for the photolysis of CF_2Cl_2 in H_2O, *Journ. Physical Chemistry*, 1975, 79, n°6, pp. 669–70.

Miramand, Y. and Guary, J. C. Association of americium-241 with adenochromes in the branchial hearts of the cephalopod *Octopus vulgaris. Marine Ecology. prog. ser.* 1981, **4**, 127–9.

Miyake, Y. Patters of radioactive contamination and future alteration of the ocean, *in* Hood, *Impingment of man on the ocean*, Wiley Interscience, 1971, pp. 569–88.

Molina, M. J. and Rawland, F. S. Some unmeasured chlorine atom reaction rates important for stratospheric modeling of chlorine atom catalysed renoval of ozone. *Journ. Physical Chemisty*, 1975, **79**, 667–9.

Moore, J. A., Gupta, B. N., Zinkl, J. G. and Vos, J. J. Postnatal effects of maternal exposure to 2, 3, 7, 8 Tetrachlorodibenzodioxin (TCDD). *Environ. Health Perspect.*, 1973, **5**, 81–85.

Moore, J. A. Toxicity of 2, 3, 7, 8 Tetrachlorodibenzo-p-dioxin, in Ramel chlorinated phenoxyacids and their dioxin. *Ecol. Bull. Stockholm*, 1978, **27**, 934–44.

Moore, N. W. and Tatton, J. O'. G. Organochlorine insecticide residues in the eggs of sea birds. *Nature*, 1965, **207**, 42–3.

Moriarty, F. *et al. Organochlorine insecticides: persistant organic pollutants* Academic Press, 1975, 302 pp.

Moriarty, F. Terrestral animals, *in* Butler, *Principles of ecotoxicology*, Wiley 1978, pp. 169–86.

Morre, J., Janin, F., Gilles, G. *et al*. Récente contamination de la chaîne alimentaire par les retombées radioactives en France. *Bull. Acad. Veter.* France, 1977, **50**, 51–8.

Morris, B. F. Petroleum: tar quantities floating in the NW Atlantic taken with a new quantitative neuston net. *Science*, 1971, **173**, 430–1.

Mosser, J. L. and Fisher, N. S. and Wurster, C. F. Polychlorinated biphenyls and DDT alter species composition in mixed culture algae. *Science*, 1972, **176**, 533–535.

Muirhead-Thomson, R. C. Impact of pesticides on aquatic invertebrates in nature. In *Pesticides and Freshwater Fauna*. Acad. Press., 1971, chap. 6, pp. 157–179.

Mullin, J. B. and Riley, J. P. Cadmium in sea water and in marine organisms and sediments. *J. Marine Res.*, 1956, **15**, 103–8.

Murozumi, M., Chow T. J., and Patterson C. Chemical concentrations of pollutants, lead aerosols, terrestrial dusts and sea salts in Greenland and antarctic snow strata. *Geochem. Cosmochem. Acta*, 1969, **33**, 1247–94.

Nash, R. G. and Woolson, C. A. Persistance of chlorinated hydrocarbons in soil. *Science*, 1967, **157**, 224–7.

Nash, T. H. The effects of air pollution on plants, particularly vascular plants. In Ferry *et al.* London University Pub., 1973, pp. 192–224.

National Academy of Sciences, *Protection against depletion of stratospheric ozone by chlorofluorocarbons*, Washington, 1979, CCN 79-57247, 392 pp.

Neel, J. V. Genetic effects of atomic bombs. *Science*, 1981, **213**,

Nelson-Smith, A. The problem of oil pollution of the sea. *Adv. Mar. Biol.*, 1970, **8**, 215–306.

Newhouse, M. L. Asbestos, *in* Lenihan and Fletcher. *The chemical environment*, Blackie, 1977, pp. 135–58.

Nishiwaki, Y., Tsunetoshi, Y. *et al.* Atmospheric contamination of industrial areas including fossil-fuel power stations, and a method of evaluating possible effects on inhabitants. Abstr. Coll. AIEA *Environmental aspects of nuclear power stations*, Vienna, 1971, pp. 247–278.

Noel-Lambot, F., Bouquegneau, J. M., Frankenne, F., and Disteche, A. Cadmium, zinc and copper accumulation in limpets (*Patella vulgaris*), from the Bristol Channel with special reference to metallothioneins. *Marine Ecology*, 1980, **2**, 81–9.

Nylander. Comptes rendus des séances de la Societé Botanique de France, 13 July 1886.

O'Connors, H. B., Wurster, C. F., Powers, C. D., Biggs, D. C., and Rowland, R. G. Polychlorinated biphenyls may alter marine trophic pathways by reducing phytoplankton size and production. *Science*, 1978, **201**, 737–9.

Oden, S. The acidity problem: an outline of concepts. *Water, Air, Soil Poll.* 1976, **6**, 137–166.

Odum, E. P. *Fundamentals of ecology*. Saunders, Philadelphia and London, 1st edn, 1959, 546 pp.

Odum, W. E., Woodwell, G. M., and Wurster, C. F. DDT residues absorbed from organic detritus by fiddler crabs. *Science*, 1969, **164**, 576-7.

Oort, A. H. The energy cycle of the earth. *Scient. Amer.* 1970, **223**, 14-23.

Ottar, B. The long-range transport of mercury and chlorinated hydrocarbons in the atmosphere, *in* proc. int. conf. *Chemistry, environment and man*, Gottlieb Duttweiler Institute, Zurich, 1981, pp. 105-24.

Peakall, D. B. Pesticides induced enzymes breakdown of steroids in birds. *Nature*, 1967, **212**, 505-506.

Peakall, D. B. Pesticides and the reproduction of birds. Scient. Amer., Avril 1970, 73-8.

Peres, J. M. and Bellan, G. Aperçu sur l'influence des pollutions sur les peuplements benthiques. Marine pollution and sea life. *Fishing News*, 1972, **12**, 3-14.

Peterlee, T. J. DDT in antarctic snow. *Nature*, 1969, **224**, 620-1.

Petroleum in the marine environment. National Academy of Sciences, Washington, 1975, 107 pp.

Pimentel, D. and Andow, D. A. Pest management and pesticides impact. *Insect. Scien. Applic.*, 1983, **5**, 141-149.

Pimentel, D. and Edwards, C. A. Pesticides and ecosystems. *Bioscience*, 1982, **32**, 595-600.

Porter, R. D. and Wiemeyer, S. N. Dieldrin and DDT: effects on sparrow hawk eggshells and reproduction. *Science*, 1969, **165**, 1999-2000.

Puiseux, Jeanne-Levain, N., Roux, F., Ribier, J., and Borghi, H. Analyse des effets du lindane, insecticide organochloré, au niveau cellulaire. *Protoplasma*, 1977, **91**, 325-41.

Raffin, J. P., Godineau, J. C., Ribier, J. *et al.* Evolution sur trois années de quelques sites littoraux de Bretagne après pollution pétrolière (*Amoco Cadiz*, 1978). *Cahiers de Biologie marine,* 1981, **XXII**, 323-48.

Ramade, F. Sur la présence d'altérations ultra-structurales dans le cerveau de *Musca domestica* après intoxication au Lindane. *C.R. Ac. Sci., Paris*, 1966, **263**, 271-4.

Ramade, F. Contribution à l'étude du mode d'action de certains insecticides de synthèse, plus particulièrement du indane, et des phénomènes de résistance à ces composés. *Ann. Inst. Nat. Agron. Paris, T V (NS)*, 1967, 268 pp.

Ramade, F. La pollution des eaux par les insecticides organochlorés et son impact pour la faune aquatique. *La Nature, Paris, Dunod*, 1968, **3404**, 441-448.

Ramade, F. La pollution par les défoliants et ses conséquences écologiques. *Courrier Nature*, 1977, **47**, 24-31.

Ramade, F. *Eléments d'ecologie. Ecologie appliquée.* McGraw-Hill, Paris, 3rd edn, 1982 (1974), 451 pp.

Ramade, F., Cosson, R., Echaubard, M., Le Bras, S., Moreteau, J. C. and Thybaud, E. Détection de la pollution des eaux en milieu agricole. *Bull. Ecol. Paris*, 1984, **15**, 1-17.

Ramade, F., Echaubard, M., Le Bras, S., Moreteau, J. C. Influence des traitements phytosanitaires sur les biocoenoses limniques. *Acta Oecol. Oecol. Aplicata*, 1963, **4**, 3-21.

Ramade, F. and Roffi, J. Action de deux insecticides, le lindane et le D.D.V.P. (Dichlorvos) sur les surrénales de la souris. *C.R. Ac. Sci., Paris*, 1976, **282**, 1067-70.

Ramel C. *et al.* Chlorinated phenoxyacids and their dioxins. *Ecological Bull. Stockholm*, 1978, **27**, 301 pp.

Rao, D. N. and Le Blanc, F. Effects of sulfur dioxide on the lichen algae with special reference to chlorophyll. *Bryologist*, 1965, no. 69.

Ratcliffe, D. A. Decrease in eggshell weight in certain birds of prey. *Nature*, 1967, **215**, 208.

Ratcliffe, D. A. Changes attributable to pesticides in egg breakage frequency and eggshell thickness in some british birds. *J. Appl. Ecol.*, 1970, **7**, 67-115.

Ratcliffe, D. *The peregrine falcon.* Calton, T. and D. Poyser, 1980, p. 212.
Reggiani, G. Anatomy of a TCDD spill: the Seveso accident. In *Hazards assessment of chemicals: current developments*, London, Academic Press, 1983, pp. 269-311.
Rice, S. D., Short, D. W., and Karinen, J. F. Comparative oil toxicity and comparative animal sensitivity, in Proc. Symp. on fate and effects of petroleum hydrocarbons *in* Wolfe, ed. *Marine ecosystems and organisms.* Pergamon Press, NY 1977, pp. 78-94.
Risebrough, R. W. Chlorinated hydrocarbons, *in* Hood, ed. *Impingment of man on the ocean.* Wiley Interscience, 1971, pp. 259-86.
Risebrough, R. W., Huggett, R. J., Griffin, J. J., and Goldberg, E. Pesticides: transatlantic movements in the northeast trades. *Science*, 1968, **159**, 1233-6.
Risebrough, R. W., Lappe, B. W., and Walter, W. Transfer of higher molecular weight chlorinated hydrocarbons to the marine environment, in Windom and Duce *Marine pollutant transfer.* Lexington Books, Toronto, 1976, pp. 261-321.
Rizebrough, R. W., Reiche, P., Peakall, D. D., Herman, S. G. and Kirven, M. N. Polychlorinated biphenyls in the global ecosystem. *Nature*, 1968, **220**, 1098-1102.
Robison, W. L. and Wilson, D. W. Modelling radiation exposures to populations from radioactivity released to the environment. AIEA Vienna, Environmental Behaviour of Radionucleides Released in the Nuclear Industry, 1973, pp. 553-70.
Roffi, J. and Ramade, F. Effet de l'intoxication à long terme *per os* par deux insecticides le Lindane et le Fenthion, sur les surrénales de la souris. *Bull. Soc. Zool. Fr.*, 1981, **106**, 167-176.
Rohwer, P. S. Relative radiological importance of environmentally released Tritium and Krypton 85. In *Environmental behaviour of radionucleides released by nuclear industry*, AIEA Vienna, 1973.
Sawicka-Kapusta, K. Roe deer antlers as bioindicators of environmental pollution in southern Poland. *Environ. Pollut.*, 1979, **A19**, 283-293.
SCEP *Man's impact on the global environment. Radioactive wastes as a function of expanding US nuclear power.* MIT Press, Cambridge Mass., 1970, p. 301.
Schmid, W. The micronucleus test. *NUT. Res.*, 1975, **31**, 9-15.
Schofield, C. *Trans. Am. Fish. Soc.*, 1965, **94**, 227.
Schulz, D. Beitrag zur allgemeinen pathòlogie des Endokardreaktionen Toxisch bedingte endokardveränderungen bei niederen Wirbeltieren (Karpfen). *Vischows Arch. A., Dtsch.*, 1973, **358**, 273-280.
Schultz-Baldes, M. and Cheng, L. Cadmium, *in Halobates micans from the central and south-Atlantic Ocean. Mar. Biol.*, 1980, **59**, 163-168.
Schupbach, M. R. Halogenierte kohlenwasserstoffe in der Nahrung, *in* proc. int. conf. *Chemistry, environment and man.* Gottlieb Duttweiler Institute, Zürich, 1981, pp. 105-124.
Schuphan, W. Der Nitrat gehalt von spinat (*Spinacea oleracea*) in beziehung zur methemoglobinemie der saüglinge. *Z. Ernährungswiss*, 1965, **5**, 207.
Scientific Committee of the White House. Restoring the quality of our environment. Washington DC, 1965.
Scope Environmental issues, n° 10, Holdgate M. W., and White G. F. eds. Wiley, 1977 242 p.
Scott, T. G. and Eschmeyer, P. H. Summaries of selected studies on wildlife pollution. Patuxtent Wildlife Research Center, US Fish and Wildlife Service. US Government Printing Office, Denver, 1980, 44-5.
Selikoff, I. J., Hammond, E. C., and Churg, J. Asbestos exposure and neoplasia. JAMA, 1968, **204**, 106-12.
Seugé J., Bluzat, R., and Rodriguez-Ruiz F. J. Effets d'un mélange d'herbicides (2, 4 D et 2, 4, 5 T): toxicité aiguë sur 4 espèces d'invertébrés limniques. *Environ. Poll.*, 1978, **16**, 87-104.
Shaw, T. L. and Brown, V. M. The toxicity of some forms of copper to rainbow trout *Water Res., Arch.* 1974, **8**, 377-382.

Sheppard, H. H. *The chemistry and toxicology of pesticides*. Minneapolis, Minn., Burgess Pub. Co., 1947, 383 pp.
Sheppard, O., Wong, W. S., Vehara, C. *et al*. Lower threshold and greater bronchomotor responsiveness of asthmatic subjects to SO_2. *Am. Rev. Resp. Dis.*, 1980, **122**, 873-8.
Shiraishi, Y. *et al*. Chromosomal aberrations and their frequencies in human leucocytes after treatment with cadmuim sulphide. *Proc. Jap. Acad. Sci.*, 1972, **48**, 133.
Sims, J. L. and Pfaender, F. K. Distribution and biomagnification of hexachlorophene in urban drainage areas. *Bull. Environ. Toxicol. Contam.*, 1975, **14**, 214-20.
Siou *et al*. Activité mutagène du benzène et du benzopyrène. *Cah. Notes Doc. INRS*, 1977, no. 59, 433-44.
Skye, E. Lichens and air pollution. *Acta Geographica Suecica*, 1968, **52**, 1-23 (Univ. Uppsala).
Smith, J. E. *Torrey Canyon pollution and marine life*, CUP, 1968, 196 pp.
Smith, R. Nuclear plant occupational exposures causing concern in industry. AEC Nucleonics week, 21 November 1974, pp. 2-3.
Smith, R. J. Utilities choke on asthma research. *Science*, 1981, **212**, 1251-54.
Snyder, N. F., Snyder, H., Lincer, J. L., and Reynolds, R. T. Organochlorine, heavy metals and the biology of North American accipiters. *Bioscience*, 1973, **23**, 300-5.
Söderlund, R. and Svensson, B. M. The global nitrogen cycle, *in* N, P, S global cycles. Ecolog. Bull. (Stockholm), 1976, **22**, 22-73.
Spinrad, B. I. The role of nuclear power in meeting world energy needs. Abstr. coll. AIEA *Environmental aspects of nuclear power stations*, Vienna, 1971, pp. 57-90.
Stebbing, A. R. D., Akesson B., Calabrese A. *et al*. The role of bioassays in marine pollution monitoring. *Rep. meet. Int. conn. Explor. Seas*, 1980, **179**, 322-32.
Steele. R. L. Effects of certain petroleum products on reproduction and growth of zygotes and juvenile stage of alga *Fucus edentatus, in Proc. Symp. on fate and effects of petroleum hydrocarbons in marine ecosystems*. Pergamon Press, New York, 1977, pp. 138-42.
Stendell, R. C., Smith P. I., Burnham K. P., and Christensen R. E. Exposure of waterfowl to lead: a nationwide survey of residues in wing bones of seven species. *US DI, Fish and Wildl. Serv., spe. rep.*, 1979, **223**, 12 pp.
Stickel, L. F. Pesticides residues in birds and Mammals. *Environ. Pollution by Pesticides*, C. A. Edwards, Plenum Press., London and NY, 1973, pp. 255-312.
Stickel, L. F., Dieter, M. P., Tait, M. D., and Hall, C. H. Ecological and physiological toxicological effects of petroleum on aquatic birds. US Fish and Wildlife Service, Biological Services Program. Rpt., 1979 FWS/OBS 79/23, 14 pp.
Stockner, J. G. and Antia, N. J. Phytoplancton adaptation to environmental stresses from toxicants, nutrients and pollutants — a warning. *J. Fish. Res. Board. Canada*, 1976, **33**, 1089-1096.
Streit, B. Uptake, accumulation and release of organic pesticides by benthic invertebrates 2. Reversible accumulation of Lindane, paraquat and 2,4 D. Arch. Hydrobiol. Stuttgart, 1979, n° 3/4 p. 349-72.
Szakvary, A. *Les résidus de produits phytosanitaires et leurs incidences possibles*. Paris, Centre de recherches Foch, 1965, 188 pp.
Tejning, S. Biological effects of methylmercury dicyanamide treated grain in the domestic fowl *Gallus gallus. Oikos*, 1967, Suppl. 8, 1-116.
Truhaut, R. Sur les dangers de cancerisation pouvant résulter de la présence de résidus de pesticides dans les aliments. *Phytriatrie-Phytopharmacie*, 1955, special no., p. 57.
Truhaut, R. Sur la prévention des risques de nocivité pouvant provenir de la présence de résidus pesticides dans les aliments végétaux. *Phytiatrie-Phytopharmacie*, 1956, 3.
Truhaut, R. Pollution de l'air. C. R. du Coll. int. de Royaumont (avril 1960) Paris, SDES.
Truhaut, R. Survey of the hazards of the chemical age. In *Abst. Int. Symp. Chemical control of the human environment*. Johannesburg, 1969, Butterworth ed., 1970, pp. 419-36.

Truhaut, R. Écotoxicologie et protection de l'environnement. Abst. Col. *Biologie et devenir de l'homme*, Sorbonne, 18-24/9/1974, pp. 101-121. MacGraw-Hill, 1976.
Truhaut, R., Chanh P. *et al.* Toxicité à long terme de la Tetrachloro 2, 3, 7, 8-p-dibenzodioxine chez le rat. Étude structurale, ultrastructurale et histochimique. *C.R. Ac. Sci., Paris*, 1974, pp. 1565-69.
Turusov, V. S., Day, N. E., Tomatis, L. *et al.* Tumors in CF1 mice exposed for six consecutive generations to DDT. *J. Nation. Can. Inst. USA*, 1973, **5**, 983-97.
Ui, J. Mercury pollution of sea and fresh water, its accumulation in water biomass. *Rev. Int. Oceanog. Med.* (Fr.), 1971, **22-3**, 79-128.
Upton, A. C. The biological effects of low-level ionizing radiation. *Scient. Amer.* 1982, **246**, 29-37.
US President's Council on Environmental Quality, 'Environmental Quality', 1971. US Government Printing Office, Washington DC.
Van Den Bosch, R. *et al. in* Huffaker, *Biological control*, Plenum Press, 1971, p. 38.
Van der Post, D. C. Some effects of silicone oil pollution on the aquatic environment. *Water Pollution Control*, 1979, 389-94.
Van Der Post, D. C. and Toerien, D. F. The retardment of algal growth in maturation ponds. *Water Research*, 1974, **8**, 593-600.
Varney, R. and MacCormac, B. Atmospheric pollutants. In MacCormac, *Introduction to the Scientific Study of Atmospheric Pollution*. Dordrecht, Reidel, 1971, pp. 8-52.
Veith, G. D., Kuehl, D. W., Leonard, E. N. *et al.* Fish, wildlife and estuaries: PCB and other organic chemical residues in fish from major watersheds of the US, 1976. *Pestic. Monitor. J.* 1979, **13**, 1-12.
Viale, D. Fréquence des accidents survenus à des cétacés sur les côtes Tyrrhéniennes. *Bull. Soc. Zool. Fr.*, 1974, **1**, 146-7.
Viale, D. Relation entre les échouages de cétacés et la pollution chimique des mers ligures et Tyrrhéniennes. FAO DOC ACMRR/MM/SC, 1976, **92**, 17 pp.
Vie Le Sage, R. Les pluies acides: un holocauste écologique. *La Recherche*, 1982, n° 131, 394-406.
Ware, D. M. and Addison, R. F. PCB residues in plankton from the gulf of Saint Laurence. *Nature*, UK, 1973, **246**, 519-21.
Weber, R. P., Coon, J. M., and Triolo, A. J. Nicotine inhibition of the metabolism of 3, 4 benzopyrene, a carcinogen in tobacco smoke. *Science*, 1974, **184**, 1081-3.
Webster, W. S. The effects of carbon monoxide on experimentally-induced atherosclerosis in the squirrel monkey. Abot. AMA Air Pollution Medical Research Conference, October 1970, New Orleans.
Weinstock, B. and Niki, H. Carbon monoxide balance in nature. *Science*, 1972, **176**, 290-2.
Westing, A. H. *Herbicides in war, the long term ecological and human consequences.* Stockholm International Peace Research Institute, London and Philadelphia, Taylor and Francis, 1984, p. 999.
Whitehead, C. C. Growth depression of broilers fed on low levels of 2, 4 dichlorophenoxyacetic acid. *Br. Poult. Sci.*, 1973, **14**, 425-7.
Wiemeyer, S. N., Spitzer, P. R., Krantz, W. C., Lamont, T. G., and Cromartie E. Effects of environmental pollutants on Connecticut and Maryland ospreys. *Journal Wildlife Manage.*, 1975, **1** 124-39.
Wilcox, K. R. Mercury levels in a sample of Michigan residents, *in* Hartung and Dinman *Environmental mercury contamination*, Ann. Arbor, 1972, pp. 82-5.
Wilson, K. W. Monitoring dose-response relationships. *Rap. Ren. Cons. Int. Explor. U. Mer.*, 1980, **179**, 333-8.
Winteringham, F. P., Lewis S. E. On the mode of action of insecticides. *Ann. Rev. Entom.*, 1959, pp. 303-18.
Wolff, P. Mercury content of mussels from West european coasts. *Mar. Pollut. Bull.*, **6**, 1975, 61-3.

Wright, A. M. and Stringer, A. The toxicity of thiabendazole, Benomyl, Methyl benzinmidazol Zyl—carbonate and Thiophanate methyl to the earthworm *Lumbricus terrestris*. *Pestic. Sci.*, 1973, **4**, 431–2.

Wurster, C. F. DDT reduces photosynthesis by marine phytoplankton. *Science*, 1968, **159**, 1474–5.

Young, D. R., MacDermott-Ehrlich D., and Heensen T. C. Sediments as source of DDT and PCB. *Mar. Pollut. Bull.*, 1977, **8**, 254–7.

Subject Index

Acetylcholine, 15
Acid rains, 75, 76, *156-159*, 161
Acrolein, 169
Adaptation, *45-48*
Adirondacks, 158
Adrenal glands, 111, 190
Adrenaline, 15, 112
Aeropollutants (*see* Atmospheric pollutants)
Aerosols (*see* Atmospheric particles)
Aflatoxins, 25, 30, 34, 43
Agent orange, 137, 139, 142, 143
Agriculture, 68, 135, 145-150
Agroecosystems (*see* Ecosystems)
Alad, 199-200
Alaska, 78, 99
Aldrin, 22, 79, 83, 84, 104
Allergies, 23, 177
Aluminium, 198-199
Amanitin, 43
Ames test, 24-26, 115, 124
Amoco Cadiz, 182, 184, 185, 190, 192
Amphetamines, 17
Antagonism, 23, 35
Antarctica, 75, 78, 95
Antibiotics, 41, 45, 69
Anticholinesterasic agents, 17
Apholate, 22
Arizona, 124
Arochlor, 92, 110
Aromatic amines, 30
Arsenic, 19, 30, 147, 176
Asbestos, 28, *178-179*
Asbestosis, *178-179*
Asthma, 178
Atlantic ocean, 55, 95, 97, 99, 183
Atmosphere, 29, 54, 71-75, *150-181*
Atmospheric circulation, 72, 73
Atmospheric particles (dusts), 70, 72-74, 99, 173-174
ATPases, 15, 114
Aziridine group, 22
Azodrin, 145

Baltic Sea, 100
Baygon, 17
Becquerel, 208
Behaviour, 19
Benomyl, 144
Benzidine, 28
Benzopyrene, 30, 152, 169, 172
BHC, 36
Bikini Atoll, 220
Bioassays, *56-58*
Biocoenoses, 81, 128, 145-146, 199
  marine, 128, 129, 228-229
Biocoenotic effects, 145, 146, 185-188
Bioconcentration, 78-86, 217-218, 226-233
Bioconcentrators, 78, 80, 97, 227-231
Biodegradability (biodegradation), 77, 185
Biogeochemical cycles, 69, 71, 173, 236
Biomass, 86, 96, 98
Biosphere, 93, 95, 98-102
Biotests, 56-57
Biotic potential, 21, 212
Bronchitis, 65, 171, 178

Cadmium, 30, 67, 124, 131-135
Caesium, 137, 65, 66, 206, 218-219, 224-232
California, 78, 102, 103, 174, 186
Canada, 25, 103, 217
Cancer, 27-29, 169, 172
  lung, 28, 172, 178, *179*, 210
Captan, 34, 144
Carbamates, 15-17
Carbaryl, 17
Carbon dioxide, 74
Carbon monoxide, 19, 41, *150-152*
Carboxyhaemoglobin, 151, 170
Carcinogenesis, *26-31*, 115, 171, 210, 214
Carcinogens (carcinogenic agents), 23, *26-31*, 115, 172, 179
Carcinomas (*see* Cancer)
Catecholamines, 15, 111-112
Chelating agents, 51-52
Chemosterilants, 22
Chicago, 122

Chlordane, 20, 79, 90
Chlorophyll, 167
Chromates, 30
Chromatography, 52–53
Chromosomes, 24, 129, 130
Cigarette, 26, 170, 179
Clear Lake, 82
Coal, 62–63, 150
Cobalt, 35
Cobalt 60, 207, 228
Colourings and dyes, 30
Copper, 51–52, 124
Cornwall, 187
Curie (Ci), 208
Cyanides, 201
Cyclamates, 30
Cyclodienes, 90, 114, 115

2,4D, 137, 139
DDA, 111
DDD, 82–83
DDE, 46–48, *108–113*, 117
DDT, 15, 21, 22, 37, 42, 53, 68, 77–80, *86–119*
  effects on birds, 105–111
  incorporation into trophic systems, 80, 86, 97–98
  on entomofauna, 103
  on fishes, 104, 105
  on invertebrates, 103, 104
  on primary production, 115–117
Defoliants, 137–146
Demoecological effects, 136–138
Detergents, 193–197, 200–201
  anionic, 193–195
  cationic, 193, 195, 201
  non-ionic, 193, 195
Detoxification, 19–20
DFP, 17
Dichlorvos (DDVP), 17
Dieldrin, 17, 79, 90, 98, 104, 109
Dioxin, 31, *140–144*
Diphenl, 21
Dispersants (of oil slicks), 194–197
Diuron, 138–139
Diversity, 146, 164
DNA, 24, 27, 207, 209, 223
Dose–response curves, *32–36*
Doses, 5–13, 48–50
  lethal, 5–13, 42, 48–50, 210–211
  sublethal, 5, 211–214
Dusts, 71–73, 150
  atmospheric, 150, *176–179*
  metallic, 173–175

Dystrophication, 181, 200

Ecological pyramid, 86
Ecophase, 44
Ecosystems, 56, 123–127, 228
  agro, 83, *135–147*, 231–233
  forest, 132
  limnic, 56, 198–201
  marine, *152–198*, 227–231
Ecotoxicology, 59
Eggshells, 107–109
Embryos, 22, 31, 43, 131, 192
Emphysema, 178
Endrin, 104
Enolase, 181
Enzymatic induction, 20, 110
EPA (Environmental Protection Agency), 55, 167
Eserine, 15
Estuary, 86
  Clyde, 94
  Thames, 128
Ethylic cirrhosis, 30
Europe, 63, 65, 90, 105, 124, 204

FAO, 132
FDA (Food and Drugs Administration), 37, 125
Fenthion, 136
Fertilizers (chemical), 68, *146–149*
Fibrosis (pulmonary), 178
Fish fry, 57, 104, 112
Fluorides, 70, *180–181*
Fluorine, 92–93, *179–181*
Fluorosis, 181
Foetus, 172
Food additive, 30, 41, 69
Food chains, 81–86, 97–102, 122, 124, 228–230
Forests, 64, 89, 132, 161, 181
France, 63, 67, 84, 105, 204, 217, 231
Freons, 93, 179
Fuels, *62–64*, 152–153, 173
  nuclear, 224–225
Fungicides, 90, *144*
  organomercurial, 119–120

Gametes, 211
Germany (FRG), 132
Glands
  adrenal, 18, 111–112
  endocrine, 18–19, 110–111
  pituitary, 110–111
  thymus, 223

Glands (*continued*)
  thyroid, 18, 111, 216, 218
Gonads, 19, 22, 199, 210, 211
Graphs log dose–probit, *7–13*
Great Britain, 61, 77, 105, 107, 158, 162–163, 172, 230.
Greenland, 99, 174

Habitats, 199
Half-life, 78, 207–208
Heptachlor, 79, 90
Herbicides, 19, 21, *137–142*
Hexachlorobenzene, 84, 94
Hexobarbitone, 20
Hiroshima, 66, 205, 210, 214
Hormetic effects, 36
Hormones, 18, 19, 21, 30, 110–111
Hydrocarbons, 56, 188–189
  halogenated, *88–119*
  pollution of the ocean by, 70, 118, *182–192*
  polycyclic, 30, 40, 152
Hydrosphere, 181–202

IC 50, 5
Idiosyncrasy, 44
India, 213
Industries, 66, 67, 119, 198, 220
Insecticides, 45, 48, 68, 135
  organochlorine, 17, 20, 89–118
  organophosphorus, 17, 42, 49, *136–137*
Iodine, 79
Iodine 131, 207, *217–218*, 221–233
Irradiation (*see* Radiation)
IRPTC, 54
Itaï itaï disease, 135
Ixtoc One, 182

Japan, 82, 125

Kerala, 213
Kidney, 20–21
Krypton 85, 75, 207, 222, *234*

Lake
  Clear (*See* Clear Lake)
  Great, 127
  Michigan, 104, 127
  oligotrophic, 138
  Ontario, 51
  St Clair, 127
  Superior, 198
Lapland, 75, 84, 85
Lapps, *84–85*, 217

LAS (linear alkane sulphonate), 193
LC 50, 5, 12–13, 105, 189, 194–195
LD 50, 5, 12–13, 36, 42, 45, 48, 200–201, 210
LT 50, 5
Lead, 41, 147, *173–175*
  alkyl, 173
  poisoning, 175
  shots, 199–200
Lindane, 8, 42, 45, 90, 96, 112, 114
Liver, 19
Longevity, 209

MAC (maximum admittable concentration), 37, 56, 101
Manchester, 175
Magnesium, 198
Manganese, 176
Mangroves, 137
Mediterranean Sea, 80, 128, 182–183, 194–197
Mercury, 20, 21, 37, 82, *119–131*
Mercury methyl, 37, 82, 120–121
Mesothelioma (pleural), 28, 179
Metallothioneins, 48
Metals (heavy), 37, 51, 70
Methaemoglobinaemia, 147
3-Methylcholanthrene, 20
Microplankton, 187
Microsomes, 20, 110
Milk contamination, 37, 84–85, 101, 102, 219
Minamata disease, 37, 82, *128–129*
Mines, 198
Monitoring of pollutants, *54–58*
Muds, 122, 197
Muscarine, 16
Mussel watch, 55, 56
Mutagenesis, 22, *23–26*
Mutagens, 22, 24, 129, 134, 141, 212–213
Mutations, 22, *23–26*, 129–130, 141, 213

Nekton, 185
Neostigmine, 16
Nerve gas, 14
Neuroendocrine system, 18
Neurotoxicity, 14–17
Neurotransmitters, 16
Neuston, 185
Nevada test site, 216
New Brunswick, 103
New Mexico, 113, 124
New Zealand, 133
Nicotine, 16, 17

Nitrates, 147–149
Nitrites, 147–148
Nitrogen oxides, 50, *152–153*, 168
Nitrosamines, 25, 30, 37, 148
Nuclear
  industry, *220–225*
  pollution, *203–235*
  power, 65, 66, 202
  reactors, 66, 75, 221–223
  reprocessing plants, 41

Oil, 63, 64, *182–193*
Oil wells, 182, 192
Olympic Bravery, 182
Ontario, 158, 161
Organochlorine, 19, 21, *89–118*
Organohalogens, 18, *88–118*
  chemical structure, 90–93
  effects on animal populations, 103–113
  effects at cellular level, 113–114
  on vertebrates, 104–113
Organomercurial compounds, 120–121
Organosilexans, 201
Oxidases, 20, 50
Ozone, 74, 154, *167–169*

Pacific Ocean, 55, 220
PAN (peroxyacyl nitrates), 50, *153–154*
Paraquat, 19, 42, 139
Parathion, 17, 49
Particles, 70, 72–74, 173, 176–179, 216
PCBS, 21, 78, 91–92, *94–110*, 116–118
Penicillin, 42
Pesticides, 68–69, 88, *135–145*
  effects on birds, 105–112
  effects on entomofauna, 103
  effects on fishes, 104–105, 139
  effects on invertebrates, 103–104, 137–138
Petroleum (*see* Oil)
pH, 75–76, 121, *156–158*
Phenacetin, 21
Phenols, 183, 200
Phenoxyacetic derivatives, *137–139*, 146
Phosphamidon, 136
Phosphates, 68, *146–147*, 200
  polyphosphates, 200
  superphosphates, *146–147*
Photo-oxydation, 187
Photosynthesis, 104, 166, 187
Phytotoxicity, 160, 180
Plankton, 79, 96, 97
Pitlochry, 156

Plants, 43, 84, 145, 154, 159–161, 180, 216, 231
Plutonium, 65, 207, 218, 221
Pollutants, 42–43, 49–50, *59–75*
  atmospheric, 71–75, *150–181*
  circulation of, *71–86*, 95–96
  classification of, *70–71*
  limnic, 82–83, 148–149, 198–201
  marine, 182–198
  monitoring of, *54–58*
  of hydrosphere, 81–102
Pollution
  atmospheric, 71–75, *150–181*
  of continental ecosystems, *135–180*
  of continental waters, 82–83, 201–202
  nuclear, 24, 85, *203–235*
  of the ocean, 98–102, *182–198*, 219–220, 227–231
  from pesticides, 93–118, *135–146*
  of the soils, 79, 83–89, 95–96, 146–147, 231–233
  sources of, 62–69
  types of, 70
Pollution and energy, 62–66
Polychlorinated biphenyls (*see* PCBS)
Polyvinyl chloride (PVC), 25, 30, *92–93*, 115
Potash (mines of Alsace), 198
Primary production, 115–117, 187–189
Probits transformation, 6–13
PWR nuclear reactors, 220–223
Pyrites, 198

Rad, 208
Radiations
  biological effects of, 209–214
  influence on biotic potential, 212
  influence on longevity, 209–211
  mutagenic effects of, 212–214
Radioactive fall-out, 85, 214–216
Radioactive wastes, 65, 66, 221–226
Radiobiology, 205–214
Radioisotopes, 206
Radionucleides, 206, 226–233
Radiosensitivity, 211–214
Radiosterilization, 212
Rains (acid), 75–76, 156–161
Rem, 208–209
Resistance (to toxics), 45–48
Retention time, 31
River Columbia, 229
River Ottawa, 125
River Rhine 198
River Rhône, 232
Ruthenium 106, 230

Sarin, 14
Scandinavia, 158
Sea
  Baltic, 100, 105, 128
  Mediterranean, 194–197
  North, 97, 100, 106, 128
  Sargasso, 68
Sedatives, 17
Sediments, 121
Seveso, 31, *142–144*
Sigmoid curve, 6
Silicosis, 178
Simazine, 22
Sludges, 197–200
Smoking habits, 38, 172
Sodium, 37, 148
Soils, 79, 83–84, 95–96, 146–147, 231–233
Soman, 14
Somatic effects, 13–21
South Carolina, 108
Steroids (sterols), 21, 110
Succession, effects on, 146
Sulphamides, 69
Sulphates, 198
Sulphur dioxide (SO_2), 65, *155–167*, 191
Sweden, 95, 123, 126–127, 140, 158
Synapses, 14–17
Syndromes
  muscarinic, 16
  nicotinic, 16
Synergism, 23, 40, 144
2,4,5T, 31, *137–141*
Tabun, 14
TBS, 193
TCDD, 140–144
Teratogenesis, 13, 31, 129, 141
Teratogenic properties, 31, 139
Tests
  Ames, 24–26
  toxicological, 4–13
Thymus gland, 223

Thyroid gland, 111, 207, 216–218
Titanium, 197
Tobacco smoke, 30, 38, *169–173*, 175
TOCP, 17
Torrey Canyon, 185–192
Toxaphene, 106–115
Toxicants, 1–5, 13–22
Toxicity, 5–51
  effects of, 13–51
Toxins, 14, 15
Tritium, 221–223
Trophic chains, 80–86, 97–102, 129, 216–218, 226–233
Trophic webs, 226–229
Tyrrhenian Sea, 197

United States, 27, 63, 66, 83, 106, 135, 148, 156, 167, 175, 204, 225
Uranium, 40, 205, 213

Vanadium, 147, 197
Vietnam War, 31, 137, 139
Vinyl chloride, 25, 28–30, 92–93
Volcanism, 71, 120, 150

Wastes
  radioactive, 65, 221–235
  mining, 197–199
Water drinking, 198
Water pollution, 125–128, 148–149, 198–202, 222, 223
Windscale reprocessing plant, 23
Wisconsin, 199
World Health Organization (WHO), 37, 125, 132, 149

X-rays, 41, 205, 208

Zinc, 51, 131, 147
Zoocoenosis, 146, 197, 200
Zooplankton, 82–83, 104, 122, 159, 188

Taxonomic Index

Abies pectinata, 161
Abra longicallus, 197
Acanthoscelides, 209
Acarina, 212
Aechmophorus occidentalis (grebe), 83
Algae, 78, 79, 185–187
Allium, 114
Allobophora caliginosa, 84
Amanita phalloides, 43
Ameirus catus, 83
Amphidimium carteri, 48
Amphora, 187
Amygdalus, 180
Anagallis arvensis, 161
Anas crecca, 199
Anas platyrhynchos (mallard), 42, 55, 109, 192
Anas rubripes, 55
Anatidae, 199
Ancylus, 139
Anguilla anguilla, 48
Annelida, 48, 194, 227, 228
Anopheles, 89
Aonidiella aurantii, 103
Ardea herodias, 127
Artemisca tridentata, 146
Arthropods, 210
Auk, 190
Avicenia, 137
Aythya affinis, 127, 199
Aythya valisineria, 199

Bacillus oligocarbophilus, 151
Bacteria, 45, 50, 77, 121, 137, 151, 153, 198
  benthic, 120
  edaphic, 50, 77
Bacterium formicum, 151
Barnacle geese, 199
Bears, 99
Bats, 113
Beauveria, 49
Birches, 135

Birds, 42, 77, 78, 83, 105–110, 123–127, 190–193, 199
Birds of prey, 105, 124
  sea, 190, 193
Bream, 134, 226
Bruguiera, 137
Bufo americanus, 84

Calopalca heppiana, 163
Cancer magister, 189
Capitella capitata, 194
Capreolus capreolus, 132
Carcharimus longimanus, 99
Carcinus moenas, 228
Cardium edule, 190
Carex scopularum, 174
Carya sp., 180
Cephalopods, 229
Cercopithecus sp., 170
Cetaceans, 197
Chiroptera, 113
Chlidonias niger, 127
Chlorella vulgaris, 122
Choristoneura fumiferana, 136
Chub, 134
Cladocera, 57, 137, 159, 200
Cladonia, 84
Cloeon sp., 138
Clostridium botulinum, 14
Clostridium welchii, 151
Coho salmon (*see Oncorhynchus*)
Coleoptera, 209
Colinus virginianus, 42
Colpomenia peregrina, 186
Columba livia, 175
Compositae, 154, 161
Copepoda, 137, 159, 200
Coregonus kiyi, 104
Coturnix coturnix (quail), 42, 43
Crangon crangon, 189
Crassostraea virginica, 80, 104
Crustaceans, 133, 137, 228, 231
Cryptogams, 84, 161–167

Cryptogams (*continued*)
   effects of SO_2 on, 161–167
Cyprinoids, 134
Cystoseira stricta, 195, 196

Daphnia, 137
Daphnia magna, 122, 201
Daphnia pulex, 212
Delphinapterus leucas, 128
Dendragapus obscurus, 136
Diatoms, 100, 159, 187, 228–230
Dicotyledons, 43, 160
Ditylenchus dipsaci, 212
Dolphin, 128
Donax vittatus, 190
Drosophila, 210, 212
Ducks, 199, 200
Dunaliella sp., 114, 116, 187

Eagle owl, 123
Earthworms, 80, 84, 144
Echinoderms, 133, 194, 227
Echinocardium cordatum, 190
Enteromorpha, 186
Ephemeroptera, 103
Esox lucius, 141
Evernia prunastri, 164

Falco peregrinus (peregrine falcon), 99, 105, *107*, 123
Fir, 161
Fish, 82, 83, 104–105, 122–127, 158, 226–227, 231
Fratercula arctica (puffin), 192
Fucus sp., 78, 186, 190
Fulmarus glacialis, 78

Galliformes, 40
Gammarus minor, 199
Gammarus oceanicus, 104
Gammarus pulex, 138
Gentians, 180
Glossosiphonia complanata, 139
Goshawk, 123
Graphis elegans, 164
Grebe, 83
Gudgeon, 228
Gulls, 190

Haenanthus catarinae, 141
Halichoerus grypus, 128
Haliethus albicilla, 102, 105
Haliethus leucocephalus, 99
Halosphera, 185

Harpacticoidea, 196
Hawk (Cooper), 124
Hemlock, 42
Herons, 127
Hypogymnia physodes, 165

Idothaea metallica, 85
Idus melanotus, 201
Insects, 47, 103, 126

Lagerstromia, 137
Lagodon rhamboides, 98
Laminaria digitata, 78, 186, 190
Laridae (gulls), 190
Latrodectus 4-guttatus, 15
Lecanora conizaeoides, 163–164
Lecidea scalaris, 164
Leistomus acanthurus, 105
Leguminosae, 161
Lepas anatifera, fascicularis, 185
Lepidoptera, 44
Lepraria incana, 164
Lichens, 84–85, *161–167*, 217
Liliacies, 80
Lobaria sp., 164
Lumbriconereis fragilis, 197
Lumbricus terrestris, 80
Lyctus sp., 212
Lymnaea palustris, 137

Macacus rhesus, 42
Macrocystis, 186
Mallard duck, 42, 55, 109, 192
Mammals, 47
Melanitta fusca, 192
Mercenaria mercenaria, 104
Methanosarcina barkerii, 151
Micropterus sp., 83, 158
Microstomus pacificus, 94
Microtus montanus (voles), 174
Molluscs, 56, 80, 97, 128, 133, 137, 190, 227–231
Monochrysis, 115
Monocotyledons, 160
Morone saxatilis, 189
Musca domestica, 45, 212
Mussels, 55, 80, 189
Myctiphidae, 94
Myotis grisescens, 113
Mytilus galloprovincialis, 80, 194

Navicula ramosissima, 187, 228
Nereis diversicolor, 48
Nerodia sipedon, 127
Nucula sulcata, 197

Oceanites oceanicus, 71
Oceanodroma leucorhoea, 78
Octopus vulgaris, 229
Oligochaeta, 80, 144
Oncorhynchus kisutch (Coho salmon), 104
Oniscus asellus, 132
Onychiurus quadriocellatus, 137
Osprey (*see Pandion*)
Ostraea, 55
Oysters, 132

Palaemonetes pugio, 189
Pandion haliethus, 105
Pannaria sp., 164
Parmelia sp., 164
Parmeliopsis ambigua, 164
Patella vulgata, 190, 192
Partridge, 124
Pecten (scallops), 133
Pelecanidae, 106–108
Pelecanus occidentalis, 106–108
Penaeus duorarum, 104
Penguins, 95
Pertusaria hemispherica, 164
Petrels, 77–78, 190
Phaeodactylum, 116–118, 186–187
Phanerogams, 97, 150–161, 210
Phasianus colchicus, 42
Physalia, 151
Picea excelsa, 161
Pigeons, 110, 175
Pike, 126, 141, 228
Pilchard, 188
Pimephales promelas, 80
Pines, 161
Pinus strobus, 160
Pinus sylvestris, 181
Plantago lanceolata, 43, 161
Porphyra umbilicalis, 186, 230
Prawn, 104
Prasinophyceae, 185
Proallariiformes, 190
Protococcus, 116
Prunus sp., 180
Pseudodiaptomus cornatus, 104
Pterodroma cahow (Bermuda petrel), 77, 78
Pterocarpus sp., 137
Pterosperma sp., 185
Puffins (*see Fraterurla*)
Puffinus creatopus, 78
Puffinus gravis, 78
Puffinus griseus, 78

Ramalina farinacea, 164
Rattus rattus, 227–228
Roach, 134
Roe deer, 132
Rhinodina roboris, 164
Rhizophora sp., 137
Rotifera, 159

Salmo fario, 199
Salmo gairdneri, 51, 105, 211
Salmonella typhimurium, 25
Salmon, 100, 103
  Coho, 104
  Kiyi, 104
Scenedesmus, 202
Scomberesox saurus, 188
Scrobicularia plana, 230
Sea birds, 190–192
Sea skater, 134
Seals, 100, 128
Skeletonema costatum, 48
Solanaceae, 154
Solenidae, 190
Somateira molissima, 192
Sparrow hawk, 105
Sphaeroma serratum, 194–195
Spongiae, 227
Spruce, 161
Stellaria media, 43, 161
Sterna hirundo, 127
Sterna sandvicencis (sandwich tern), 106
Sticta limbata, 164
Swan, 199

Tadarida brasiliensis, 113
Teleostei, 141
Terns, 106, 127
Thalassiosira pseudonana, 116–117
Thamnophis sirtalis, 84
Thunnus albacares, 99
Trachurus trachurus, 128
Tribolium, 209
Trout (*see Salmo*)
Tubifex sp., 137
Tunicata, 227

Uca pugnax, 164
Usnea sp. (fiddler crab), 97

Vole (*see Microtus*)

Woodlouse, 132

Xanthoria parietina, 163